Abundant Agriculture

Smartcultures enable superior
Nutrition and Yields from Regenerated Fields

Mark R. Edwards

The Green Algae Strategy Series

www.AlgaeAlliance.com

www.AlgaeCompetition.com

Key words.

Food	Biofertilizer	Sustainability	Fertilizer
Water	Nutrient recovery	Ecosystems	Hunger
Regenerative	Fossil free food	Smartcultures	Poverty
Agriculture	Climate change	Energy	Drought
Aquifers	Nutrient deficiency	Environment	Soil
Soil crust	Global awareness	Algaculture	Malnutrition
Genetics	Renewable energy	Biotechnology	Pollution
Microalgae	Industrial farming	Micronutrients	Algae

ISBN 1456368125

ISBN 13 9781456368128

Abundant Agriculture is the seventh book in The Green Algae Strategy Series.

Dedication

To Ann Ewen, who enriches my life with an abundance of joy, music, gardens, friendship, and great food.

To Sarah Edwards who finishes grace before meals with, "Please God, bless this food and help people who don't have food get some."

To growers who desire to leave every field better than they found it.

The Green Algae Strategy Series

By Mark R. Edwards

The Green Algae Strategy Series focuses on creating Sustainable and Affordable Food and Energy – "SAFE" production. **The Green Algae Strategy Series** are available for free downloading in color PDF for students, teachers, and food and energy policy leaders at http://GreenIndependence.org. These SAFE production books are used in schools and colleges globally for courses in sustainability, engineering, business, political science, social entrepreneurship, food, water, energy, ecology, environment, and world future.

BioWar I: Why Battles Over Food and Fuel Lead to World Hunger, 2007. BioWar I, where food is burned for fuel, must be ended by withdrawal – not of soldiers, but of damaging agricultural subsidies.

Green Algae Strategy: Engineer Sustainable Food and Fuel, 2008. Algae offer solutions for sustainable and affordable food and energy because algae are the most productive biomass source on Earth. *Best Science Book* **– 2009, Independent Publisher Awards**.

Green Solar Gardens: Algae's Promise to End Hunger, 2009. Algaculture in small but beautiful solar gardens and algae microfarms distributed globally will enable SAFE production locally.

Crash! The Demise of Fossil Foods and the Rise of Abundance, 2010. Traditional fossil-based agriculture sits precariously on a foundation of unsustainable fossil resources that will become unaffordable and then will run out. Abundant agriculture is sustainable because it uses plentiful inputs that are cheap and will not run out.

Smartcultures: **Nature's tiny Genius – Algae – Reverses Environmental Pollution and Regenerates Degraded Ecosystems**, 2011. Farmers may recycle farm wastes to their fields using algae microfarms. Smartcultures give higher yields by providing bioavailable nutrients at just the right time. Farmers save money by reducing nearly all input costs by 40%. Smartcultures reduce ecological pollution by 90%.

Contents

Abundant Agriculture –
Sustainable and Affordable Food and Energy

Modern agriculture over consumes and wastes fossil resources that will be gone in the near future when our children will desperately need them. When the first fossil resource required by modern agriculture runs out or becomes unaffordable to farmers in a few decades, our food supply will crash. Despite increases in food productivity, over one billion people are malnourished and hungry today because they lack access to food or the inputs to produce good food. Over three million are food insecure.

Our society depends on an industrial food production system that:

- Consumes 20% of its fossil fuels.
- Creates 36% of air and 80% of water pollution.
- Exploits 80% of available fresh water to grow crops.
- Systemically extracts, erodes and exhausts its fields.
- Applies millions of tons of chemical poisons annually.
- Erodes to massive dead zones that kill all aquatic life.
- Applies 30% of its grid energy to move irrigation water.
- Extracts over 200 million tons of chemical fertilizers each year.
- Imposes illness and premature death from ecological pollution.

Abundance offers an alternative food production method that uses minimal or no fossil resources consumed by industrial agriculture. Growers who practice abundant agriculture cultivate algae and diverse microbial communities in 360 microfarms as they produce food, feed, and other forms of energy using only plentiful resources that are free or surplus and will not run out – sunshine, CO_2 and wastewater. Microfarmers close the loop as they recover, recycle, and reuse energy and nutrients from farm waste streams. Abundance enables sustainable and affordable food and energy production.

Diffusing abundance methods globally will enable hungry people to produce sustainable and affordable food and energy for the needs of their family and community locally.

Preface

The prior seven books in the *Green Algae Strategy Series* address sustainable and affordable food and energy, (SAFE) production. The series outlines the path for algae production, harvest, and coproduct extraction. Algaculture can deliver SAFE production but will fail in widespread adoption unless we include those with the most knowledge of growing crops – farmers.

Abundant Agriculture creates a bridge from industrial agriculture to abundance by engaging modern farmers of every crop. Farmers practicing abundance grow algae and other microorganisms to recycle the farm waste stream. Farmers may harvest algae to recover all the valuable coproducts to use or to sell for food, feed, nutrients, fuel, fertilizer, pollution solutions, nutraceuticals, or fine medicines.

Some abundance farmers will use smartcultures to recover the farm waste stream and skip the harvest step. Smartcultures "grow and flow" the algae culture back to their fields as algae provides bioavailable, (immediately usable) crop nutrients by special delivery. Smartcultures enable farmers to:

- Increase their existing crop productivity 30%, while enhancing produce nutrients, taste, texture, and color.
- Recycle their waste streams and reduce farming costs by 30%.
- Reduce the use of agricultural poisons and pollution by 90%.

Three years ago, abundance was only a dream. Now we have field experience with results exceeding our expectations.

> Abundance mimics nature by concentrating bioavailable nutrients in algae to deliver targeted nutrients to our crops. Abundance is the most significant breakthrough in agriculture I've seen in 40 years of farming. Improving our crop quality and yields, while reducing input costs and natural resources, is a game changer for every farmer.
>
> **Scott**, Farms Manager
> for a large multinational food company

Abundance

Abundant agriculture offers a sustainable future supply but does not yet exist as an industry. A few growers are using elements of abundant agriculture, but no infrastructure exists yet to support diffusion and adoption.

Abundance represents an emerging industry. We probably have over 90% of the practice yet to learn. However, since culturing algae in the farm or other waste stream parallels growing conventional crops, the learning curve for farmers does not represent a steep slope. The incredible ingenuity and natural inventiveness that is core to our farming culture will lead to abundance enhancements.

Looking back on industrial agriculture, before the adoption of abundance methods, historians will ask why policy leaders thought the fossil resources would never run out or become too expensive. Why did political leaders fail to act to change the food supply to make it healthy rather than providing non-nutritious calories? Why did our children not act to preserve their precious, irreplaceable natural resources their own parents were frittering away? Why did engineers not figure out efficient, non-fossil methods to recycle farm and municipal waste nutrients? Why did our government fail to act to protect its citizens from the substantial health impacts of agricultural pollution? Why did someone not act to enable distributed food production with plentiful inputs available to everyone?

While some debate whether human actions caused global warming, there is no ambiguity about the cause of natural resource depletion. Our generation imposed that severe penalty on our children. We over-consumed and wasted those precious resources in order to give ourselves cheap food. When our children cannot afford good food, they may not agree that our food was cheap.

We must take action now to improve the quality and security of our food supply by reengineering the foundation of our food production system. We need a focus on sustainable, fossil-free, healthy food; with inputs available and affordable to all – abundance.

Mark R. Edwards
January, 2011.

Chapter 1. Is Modern Agriculture Sustainable?

A field becomes exhausted by constant tillage.
    — **Ovid** (Publius Ovidius Naso), Are Amatoria (III, 82*)*

The recent 60 year experiment with industrial agriculture to produce cheap food has grown more food by leveraging nonrenewable (fossil) natural resources. Our fossil food supply depends on massive amounts of these precious resources. The natural resource leverage factor is probably twice the size of the credit leverage that led to bursting the housing bubble. Unfortunately, when the first of the natural resources on which fossil foods depend, become unavailable or unaffordable to farmers locally, food crops will fail. The food bubble will burst.

The political response to the housing meltdown was to print more money. When the food bubble bursts, we will not be able to gin up more fossil resources because they will be gone. Modern extractive agriculture will very likely cause our food supply to crash in the next generation due to overconsumption and waste.

The sustainability scorecard, Table 1.1, identifies critical factors for our food supply. The scorecard covers three key areas: health and food security, economics, and access to good food.

1

Table 1.1 Food Supply Sustainability Scorecard
10 = Highly sustainable

Health and food security

1. **Nutrition and health** – provides a diversity of affordable, healthy and tasty foods that combat disease as well as nutritional, and vitamin deficiencies.

2. **Food security** – enables access to good food for everyone at all times.

3. **Climate chaos** – produces good food independent of weather or climate-related events.

4. **Risk – economic, physical and health** – moderates economic, physical and health risks from growing and consuming food.

Economics and overconsumption

5. **Costs** – moderates rising costs of food inputs.

6. **Fossil resource consumption** – uses minimal fertile soil, fresh water, fossil fuels, chemical fertilizers or fossil agricultural chemicals.

7. **Ecologically positive** – systemically enhances soil, water, and air during and after use.

8. **Transportation** – minimizes transportation energy and cost by enabling food production local to consumers.

Social – Access to food

9. **Social equity** – enables everyone to have access to good food or the inputs to grow food.

10. **Hunger and malnutrition** – mitigates hunger, malnutrition and nutritional, and vitamin deficiencies.

Is Modern Agriculture Sustainable?

Industrial agriculture

The designers of industrial agriculture in 1950 began building a new food system with the intention of feeding a population of 2.5 billion people, about 34% (850,000) of whom faced malnutrition and hunger. In 2010, world population topped 7 billion, with slightly over 1 billion malnourished and hungry, while another 2 billion are food insecure.

When the Green Agricultural Revolution began, fossil fuels and resources were as cheap and plentiful as buffalo had been a century earlier, on the prairies. The major difference was that consumption of fossil resources is not as visible as buffalo. Evidently, no one pointed out the design flaw of basing our food supply on fossil natural resources with limited supply, Figure 1.1. Better living through fossil fuels, irrigation, inorganic fertilizers, and agricultural chemicals was the mantra that drove the Green Revolution.

Figure 1.1 Supply and Demand for Fossil Natural Resources

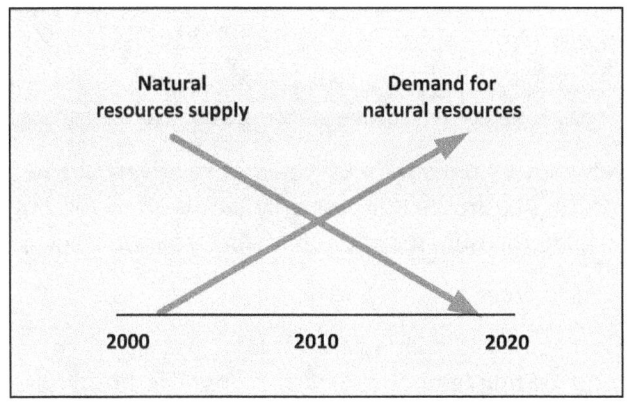

No one seemed to anticipate the reward system that paid farmers for weight would produce crops with more water, fewer nutrients, less color, and weaker taste. Few would have guessed that subsidies specifically designed to help family farmers would be hijacked by wealthy farmers and large agribusinesses.

Consider the consumption and pollution for each acre of corn, Figure 1.2. Added to these resource costs are the substantial equipment, refining, packaging, storage and transportation required for industrial food production. American farmers planted 94 million acres of corn in 2010 with 44 million acres consumed for ethanol.

Figure 1.2 Fossil Resource Consumption – One Acre of Corn

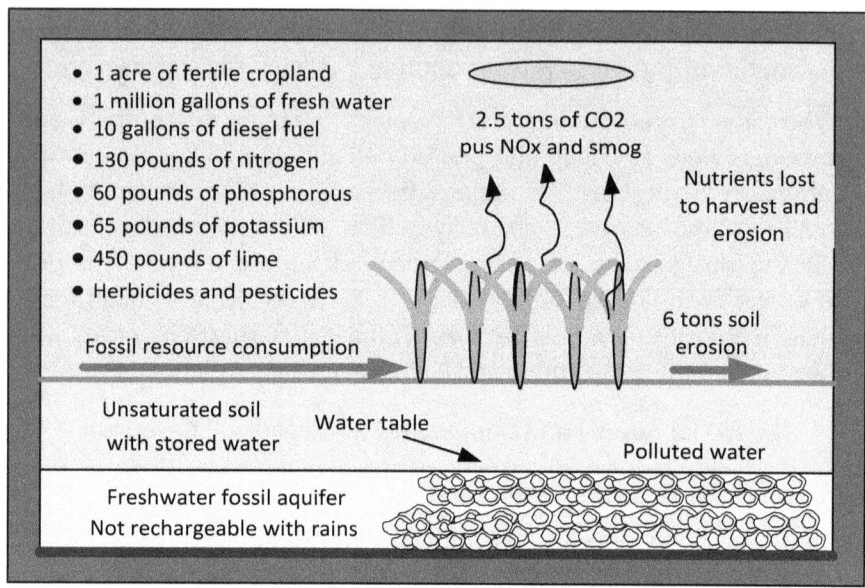

The Green Revolution did create 60 years of relatively cheap food but the concentration in production not only jeopardized the entire food supply but tripled the number of people who are food insecure.

Traditional agriculture

Agriculture consumes 12 billion acres of land globally, with about 70% in pasture and grazing land and 30% in crops. The husbandry of feed and food plants enabled humans to diversify their diets and spend less time ensuring a food supply. For most of farming history, farmers cultivated their fields using organic production methods as they:

- Grew a rich diversity of plants (called heritage varieties today).
- Planted cover crops to replace soil organics and nutrients, hold soil moisture, and minimize erosion.
- Rotated crops to improve soil fertility and minimize pests.

- Applied carbon, crop residues, animal, and green manure (plant material) for fertilization to replenish soil nutrients.

For roughly 11,000 years, agriculture was environmentally benign because farmers relied on natural ecological processes. Crop residues were incorporated into the soil or fed to livestock. Manure was returned to fields where it recycled nutrients and soil organics. Typical mixed production farms with crops and animals were closed, stable, and sustainable ecological systems that generated few external impacts. The reason traditional agriculture succeeded for multiple millennia was that mixed production of animals and food conserved vital nutrients.

Agriculture improved human societies but its Achilles' heel was that food production was dependent on considerable labor, good weather, fertile soils, sufficient freshwater delivered at just the right time and the avoidance of pests. When weather, soils or water failed, communities and sometimes entire civilizations, perished. Before the last people died from community starvation, war and illness decimated the population. History will repeat community starvation as regions and countries run out of critical inputs for modern food production.

A food supply based on fossil resources will succeed only until the first of 21 critical fossil resources are available and delivered on time, Figure 1.3. Plants can neither talk nor move, and are completely dependent on growers to supply their needs. Growers that find only one key resource unavailable or unaffordable must watch their crops wither and die.

Elizabeth Kolbert in *Field Notes from a Catastrophe* chronicled a list of sophisticated cultures that sustained themselves for hundreds of years and then crashed due to overconsumption of critical resources and climate change, primarily multiyear droughts, such as:

- Tiwanaku, Lake Titicaca in the Andes – crash: A.D. 1100, drought
- Classic Mayan civilization – crash: A.D. 800, drought
- Old Kingdom of Egypt – crash: 2,200 B.C., drought
- Akkadian empire – crash: 2,200 B.C., drought

Abundant Agriculture

Jared Diamond in *Collapse: How Societies Choose to Fail or Succeed* describes similar eco-meltdowns that caused the Anasazi of the U.S. Southwest and the Viking colonies of Greenland to crash. He shows how patterns of population growth combined with drought, over-farming, and destruction of natural resources leads to deforestation, erosion, and starvation.

Figure 1.3. Fossil Inputs needed for Modern Food Production

Primary inputs		Micronutrients – trace amounts		
1.	Fertile soil	9.	Carbon	(C)
2.	Freshwater	10.	Oxygen	(O)
3.	Fossil fuel	11.	Hydrogen	(H)
4.	Fine seeds	12.	Sulfur	(S)
		13.	Magnesium	(Mg)
Macronutrients		14.	Boron	(B)
– large amounts		15.	Copper	(Cu)
5.	Nitrogen (N)	16.	Chlorine	(Cl)
6.	Phosphorus (P)	17.	Iron	(Fe)
7.	Potassium (K)	18.	Molybdenum	(Mo)
8.	Calcium (Ca)	19.	Manganese	(Mn)
		20.	Nickel	(Ni)
Note: Nutrients must be in		21.	Zinc	(Zn)
a form usable by plants.				

Pundants are fond of saying the food surpluses we enjoy today will continue. Few of those experts have done the math. Even agricultural economists often fail to see the food sustainability issues because their models typically do not account for soil degradation, freshwater scarcity, peak oil, fertilizer availability, ecological pollution or the impacts of agricultural pollution on human and animal health. Their models are typically blind to climate chaos, which makes them practically meaningless.

Is Modern Agriculture Sustainable?

While many people hope that technology will again save human civilizations from the Malthusian trap, where population increases faster than food production, the food system sustainability and performance score card does not provide an optimistic assessment.

New technologies will require more of the natural resources that are already in short supply; most notably energy and freshwater. Promised "climate ready seeds" are at least a decade away. Even if genetically engineered, (GE) seeds enable additional production; the crops are likely to be even more vulnerable to pest vectors. Current GE crops require significantly more water, cultivation, fertilizer, and agricultural poisons than heritage crops because they cannot compete on their own with natural plants.

GE crops make the food sustainability scorecard worse because they add to the velocity of resource consumption. The human and animal health impacts of GE foods will take decades to determine. The cost of GE seed R&D, plus expensive research on health impacts will push the price of seeds beyond the reach of many farmers. Many farmers today cannot afford GE seeds.

Food sustainability

A sustainable food supply system should consistently provide affordable, healthy food for at least seven generations into the future, promote human health and vitality as well as provide food reliable security for everyone. Currently, modern agriculture provides neither affordable healthy food nor food security.

Table 1.2 Modern Fossil Food Sustainability

Nutrition and food security
1. **Nutrition and health**

Rewards drive behavior and farmers are paid based on yield. Modern food production optimizes yield (weight) in lieu of nutrition, taste or quality. Consequently, farmers add tons of nitrogen, (N) phosphorus, (P), and potassium (K) fertilizer that inflates produce with water weight. Decades of constant nutrient extraction from the soil without sufficient replenishment combined with massive amounts of chemical fertilizers produces nutrient dilution. Each year consumers receive fewer nutrients per calorie. Today consumers must eat eight field tomatoes to receive the nutrition of one tomato in 1960. Few people eat eight tomatoes. Unfortunately, when nutrients, especially micronutrients and trace minerals are not present in produce; taste, texture and color diminish too. Produce with diminished micronutrients leaves consumers with nutrient deficiencies such as calcium, manganese, copper and zinc.

Decades of nutrient depletion combined with refining methods have created foods with empty, non-nutritious calories and an obesity epidemic. Cheap foods that the poor can afford are high-fat refined foods that often cost $1/5^{th}$ as much as healthy, fresh, whole foods. Michael Pollan and others have pointed to the cheap, high-fat corn sugar that pervades our food supply causes severe health problems and shifts costs from food to health.

Poor people typically lack access to healthy food. The Centers for Disease Control reported that one out of three Americans children born after the year 2000 will contract diabetes; predominantly due to poor diet. Obesity and diabetes put our children at risk for a litany of diseases that cause our children to miss out on abundant life, ruin family lifestyles, and often leads to premature death. Diabetic children often experience chronic fatigue, mental confusion and learning disabilities. Unless major changes are made in our food supply, the social cost of the diabetes epidemic alone will destroy the American education and health systems as well as our economy.

The $43 billion a year in US ecological damage caused by agricultural pollution is not paid by farmers but by society. Industrial agriculture pollutes air, soil and water with greenhouse gases, dust, agricultural chemicals and poisons. Ecosystems polluted by agricultural runoff cause developmental disabilities and often death for fish, other aquatic creatures and water fowl. Air and water borne agricultural pollution causes significant health problems that often lead to disability and premature death for farmers and people living in rural communities. Trains from the city of Chotia Khurd in northern India are now called cancer trains because so many people in the farming villages must travel from their homes to the city for cancer treatments

2. Food security

In spite of productivity increases, over 40 countries experienced food riots in 2008 due to food insecurity; the availability and affordability of food. Over 50 world leaders called for a doubling of food production in the coming decades. Their speeches may have been pure rhetoric because food production may have already peaked along with the fossil resources on which our food depends.

Concentration of food production by a relatively few wealthy farmers and large agribusinesses amplifies food insecurity. Over 90% of modern farm production cultivates only eight crops, creating monocultures that are highly vulnerable to invasion from disease, weather or pest vectors.

In the richest country in the world blessed with extraordinary cropland, climate and natural resources, over 60 million Americans receive food support – food stamps or school lunches – because they are hungry. Those on Food stamps provide $1 for each meal, which typically goes to buy the cheapest food, which deliver empty calories and obesity. Societies lose twice from the malnutrition that accompanies food insecurity: dull witted children who have trouble learning and the medical costs of obesity and related maladies, including diabetes and heart disease.

3. Climate chaos

Food crops have been hybridized over 11,000 years to produce under stable climate conditions. Climate chaos imposes severe temperature spikes, more fierce storms fueled by warmer oceans, floods and extended droughts. Crops are sensitive to amplified heat from the higher morning lows and daily highs. Rising oceans and salt invasion not only destroy millions of hectors of crops but often the non-replaceable cropland too.

Warmer winters, later fall and earlier spring enable pests to survive over the winter. Pests begin their relentless attack on crops earlier in the plant life cycle when the plants are most vulnerable. Bark beetle infestations have ravaged forests from Alaska to the Southwest and threaten to decimate commercial forestry.

Higher temperatures also diminish crop strength, sapping their defenses. High winds twist crops so they are not harvestable and winds remove millions of tons of precious topsoil. More heat, humidity and rainfall will enable mosquitoes, ticks, and other parasites, and disease carriers to spread to areas where they did not exist previously. These tropical and subtropical disease carriers will infect populations that have not built up resistance to them. Malaria, cholera and other diseases are now being seen where they were not detected previously. After a 75-year absence, dengue fever returned to the United States in 2009 and may spread to 28 states, according to a Natural Resources Defense Council study.

4. Risk – economic, physical, and health

Industrial agriculture forces farmers to bear substantial risk of crop failure as well as physical injury and health risk. Farmers must invest in all the food production inputs, supply their labor and then wait an entire growing season, often 120 days, for their crop to produce. Unfortunately, too often crops fail due to weather or pests, which may leave the farmer bankrupt.

Farmers face substantial physical risk because farm work requires long hours of fatiguing physical labor and work around dangerous mechanical equipment. Farming, along with fishing and forestry, are the vocations with the highest probability of premature disability or death. Prolonged exposure to agricultural poisons often imposes health problems on farmers, as well as their neighboring community.

Economics and overconsumption

5. Costs

Continually rising production costs alone make modern agriculture unsustainable. Most family farmers have been pushed off their land by large farms and corporate agribusinesses. Larger farms can absorb more risk and cost variability than small farms. Every year, more of the largest food companies outsource food production to shift the risk and costs to independent operators. The massive consumption of nonrenewable crop inputs will accelerate price increases as natural resources become increasingly scarce. Competition for scarce crop inputs will be amplified by countries such as China and Saudi Arabia that are buying up available mines.

6. Fossil resource consumption

Our planet has insufficient remaining fossil resources to support the systemically increasing appetite of industrial agriculture. The five critical fossil resources needed for food production include: fertile soils, freshwater, fossil fuels, fertilizers, and fossil agricultural chemicals. Unfortunately, each these natural resources will selectively become unaffordable or unavailable to many farmers within the current generation.

Globally, 30% of the fertile soils have been abandoned over the last 30 years because they were worn out from constant extraction and erosion. Half of the remaining soils are so degraded farmers must add twice as much fertilizer, and more water, yet still see their crop yields diminishing.

In the Western US, agriculture uses 80% of the available freshwater and 30% of the grid energy goes to transporting irrigation water. Thirsty cities are buying up farmer's water rights, which mean fertile croplands will revert to prairie or desert. Large rivers, lakes, reservoirs, and groundwater wells are going dry on every food growing continent. For centuries, Beijing had over 1,000 lakes but today only a few remain. China is considering a long pipeline, not for food production but to supply municipal water for major cities. Rapidly expanding deserts make the future of food production in China problematic. Some people hope desalination will solve the water problem but desalination costs 10 times too much for use in agriculture and consumes too much energy.

Modern food production depends on massive government subsidies for storing and moving irrigation water. When water subsidies end because governments cannot afford them, so will food production because over 60% of cropland depends on irrigation. Global warming has already expanded both the geography and intensity of irrigation. Irrigation wastes water as flood, furrow and sprinkler irrigation often lose over 50% of the water before it reaches the crop.

Modern agriculture uses 23 times more fossil fuels today to produce each pound of food than in 1950. A food supply leveraged on fossil fuels can survive only as long as those fuels are plentiful and cheap. Every step in crop production – seeds, cultivation, fertilizer, and chemical applications, harvest, transportation, and food processing – use tremendous amounts of fuel. Food production consumes about 20% of the fuel used in the US. Each year, American farmers apply 13 million nutrient tons of N, 4.3 million tons of P, and 4.6 million tons of potassium, as well as other elements and chemicals. Open cycle farming loses about half these fossil nutrients to harvest and most the rest to erosion.

Fertilizers require massive amounts of fossil fuels for mining, manufacture, distribution, and application. Roughly 90% of the cost of nitrogen fertilizer comes from the cost of the natural gas used to produce it. Fertilizers prices, such as P have increased 700% in a recent 14 month period. Crops accept no synthetic substitutes for the vital P fertilizer, which jeopardizes the entire food production system because only five countries control the P mines. Some experts predict the few P mines globally will be depleted within 30 years. Without P, crops cannot perform photosynthesis and they lack the energy to germinate and grow.

Agricutltural chemicals such as pesticides, herbicides and fungicides consume additional fossil energy and depend on non-renewable elements. Resistance to agricultural poisons has become a major problem. Farmers lost about 7% of their crops to pests in 1950, before these chemicals were available. Today farmers lose about 14% of their crop to pests. Only about 1% of a pesticide is absorbed by the plant, which leaves tons of poisons to erode into wetlands, waterways, estuaries, and oceans.

7. Ecologically positive

Fossil agricultural destroys itself by continually extracting the precious nutrients and organics from the fields. Every harvest depletes about half the nutrient load in the soil. Many farmers only replenish the macronutrients, causing nutrient depletion in our foods. Cultivation breaks up the topsoil and leaves fields highly vulnerable to erosion from wind and water. Erosion carries away not only the precious nutrients and trace elements but the irreplaceable soil organic matter. Worn out soil often lacks the organic matter to hold water or the nutrients plants need and must be abandoned.

Agricultural chemicals poison our fields, wetlands, lakes, estuaries and oceans and create severe health problems and death for fish, fowl, and aquatic creatures, as well as human societies.

8. Transportation

In cities, where a majority of food is consumed, transportation represents 50% or more of food cost. The food supply chain uses massive amounts of energy because the current food production geography is distant from cities and population centers. Cities are expected to house 75% of the population by 2050, magnifying transportation costs. Without heavy subsidies for fuels, most countries and cities would face food insecurity immediately.

Social – Access to food

9. Social equity

Modern agriculture creates social inequity because those with the most need lack access to food and the inputs to grow food. Industrial agriculture, with its substitution of capital and fossil fuels for labor, has replaced family farms with corporate agribusinesses that concentrate food production in the hands of a few. The 322,000 principal US farm operators, 0.001% of the US population, produce 90% of all foods consumed in the US, plus another 11% for export.

13

Crop subsidies were designed to support family farms. Unfortunately, US Congress has bowed to the farm lobby led by a relatively few wealthy farmers and agribusinesses. Over 90% of subsidies flow to only 10% of the largest farms. Recent subsidies for biofuels, notably corn ethanol, have followed the same pattern and reward primarily the wealthy. The large farmers use the subsidies to buy smaller farms; amplifying social inequity.

10. Hunger and malnutrition

While the green agricultural revolution doubled food production, the number of people severely hungry or malnourished has more than doubled since 1950 – to over 2 billion. Nutrient deficiencies such as vitamin A, which causes blindness, iodine which causes stunting and mental retardation, and vitamin D, which causes diabetes, plague over half the children as well as many adults in over 50 countries. Modern genetically engineered monocultures are poor sources of vital nutrients because consumers receive more empty calories but fewer vital nutrients with each bite of their produce.

Our food production system has not improved the availability of nutritious food for many hungry people. UNECEF reports that a child dies from malnutrition every 3.6 seconds.

Adapted from *Crash! The Demise of fossil Foods and the Rise of Abundance*, (Edwards, 2009), which describes natural resource depletion is described in detail.

Sustainability scorecard

Our current food production system fails to support good health or food security, over-consumes and depletes our children's natural resources and leaves billions of people without access to good food and the nutrients needed for a healthy life.

Another 3 billion people are expected to be added to those already sharing our small planet by 2050. Does industrial agriculture have the capcity to support another 3 billion hungry human consumers? What if modern agriculture has already peaked, like fossil fuel supplies? How will our children grow sufficient food?

14

Is Modern Agriculture Sustainable?

The grades posted reflect the mean scores from seven convenience samples given by community groups. These scores may not have precision but the main take away is that modern agriculture the produces fossil foods flunks critical sustainability tests.

Table 1.3 Industrial Food Supply Sustainability Scorecard
1 = Not sustainable; 10 = Highly sustainable

Health and food security	Grade /10
1. Nutrition and health	1
2. Food security	3
3. Climate chaos	1
4. Risk – economic, physical and health	2
Economics and overconsumption	
5. Costs	1
6. Fossil resource consumption	1
7. Ecologically positive	1
8. Transportation	1
Social – Access to food	
9. Social equity	1
10. Hunger and malnutrition	1
Total	13/100

If a foreign country imposed these severe health, food security, resource extraction, pollution and costs on the US, we would declare war on that country. Incredibly, we have inflicted these costs on ourselves, our ecosystems, and our children.

Organic food production

Organic food production creates only modest change to the food sustainability scorecard. Most organic foods today are produced using large-scale industrial food models, due largely to food supply chain demands. The big food companies and food retailers want large volumes of organically labeled food. Large farms often have industrial and organic fields growing side-by-side, and use the same farm equipment on both. The organic fields receive non-synthetic chemical inputs, to the degree possible.

Organic food production scores slightly higher than industrial foods on health on nutrition, dietary health, and food security. The avoidance chemical fertilizers and agricultural poisons increase the score on physical and health risk. Some organic production scores higher on transportation cost but that represents only the tiny segment associated with local farmers' markets. Organic food production is just as vulnerable as fossil foods to climate chaos, economic risk, and the economics and overconsumption factors. In order to produce each calorie of produce, organic growers may use more cropland, freshwater, and fossil fuels than industrial farmers. Of course, they probably use less chemical fertilizers, and agricultural chemicals.

Organic farmers may reduce soil degradation by using fewer chemicals. However, organic growers have increased need to cultivate the soil to physically eliminate competing weeds and other pests. Cultivation kills the soil microbes, disrupts the soil structure, and increases soil erosion. Organic fertilizers must be plowed into the soil or the nitrogen will volatize rather than providing crop nutrients.

Organic farming may provide a little more social equity for small farmers but the labor and input costs are often beyond the means of many. Organic production has not made progress in fighting global hunger and malnutrition. Organic production may score 26/100 on the sustainability scorecard.

Our food supply needs to change – quickly. Reengineering our food system on a sustainable foundation of non-fossil inputs and ecosystem cleaning and regeneration may provide a solution.

Chapter 2. What is Abundant Agriculture?

I have come that they may have life and have it to the full. This life to the full is abundant life. – John 10:10

Abundant life is possible only with sufficient good food. Our society needs a new, supplementary method for food production built on a sustainable foundation that presents minimal risk and low costs for farmers. Food production should consume minimal or no fossil resources and produce excellent food without ecological pollution. The new food system should grow affordable food consistently, independent of climate, weather, geography or politics and require minimal transportation energy and cost.

The new food system should be inclusive of all people enabling access to good food or the inputs for growing food. The food system should provide solutions for hunger and malnutrition in inner-cities, barrios, slums, suburbs, rural areas, and small villages. The food produced should include the full set of the vitamins, minerals, and antioxidants essential for healthy growth and development.

Abundant agriculture – abundance – mimics nature as it leverages the robust characteristics of nature's first, simplest, fastest growing, and most resource efficient food system – algae. Abundance benefits from photosynthesis which captures and stores solar energy in energetic plant bonds in the most bountiful and beautiful plant on Earth.

Growers practicing abundance are essentially green solar farmers as they transform solar energy to rich plant biomass. The green biomass concentrates energy in chemical bonds that are portable and may be transformed to many forms of energy.

While algae are nutritious and tasty, the first generation abundance growers probably will cultivate algae primarily to feed fish, fowl, animals or their crops, Figure 2.1. Fish farmers can flow part of the culture each day to another tank and grow crustaceans or finfish. Other farmers may recycle their farm's waste nutrients and use the algae culture to supply the fertilizer to grow vegetables and fruits hydroponically or crops in their fields.

Figure 2.1 Algae used for Food and Feed

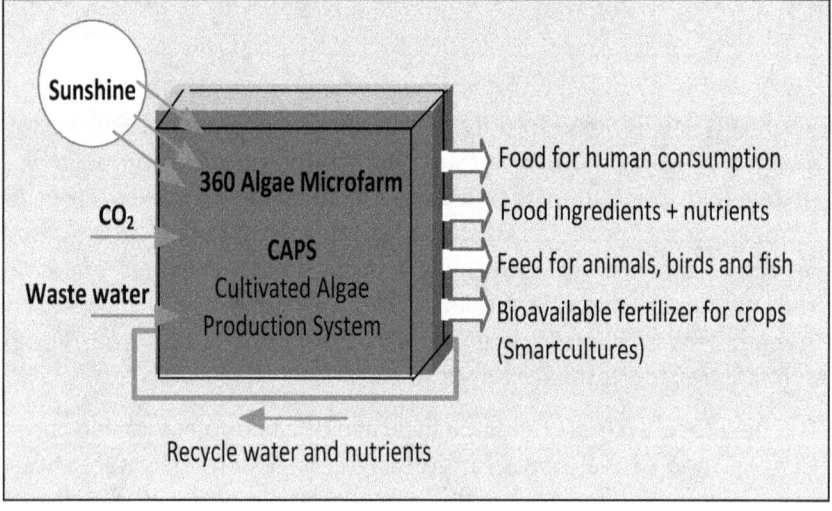

Growers use 360 algae microfarms to cultivate microorganisms that scale to any size. The footprint of microfarm production on each industrial farm will vary from 0.1 to 400 acres based on the available waste stream and farm needs. Dairy and animal farms may have 4 to 40 acres dedicated to the recovery and reuse of manure and field crop residues. Some farmers may import waste from neighbors or municipal waste recovery operations. Farmers that lack sufficient waste locally can source their CO_2 and nutrients from market sources.

What is Abundant Agriculture?

Algae

The foundation for abundance relies on a plant that used 3.7 billion years of evolution wisely to develop strategies to:

- Adapt to acute searing and freezing temperature spikes.
- Survive the extremely hot temperatures of early Earth.
- Grow in ocean, brine, saline, waste, and fresh water.
- Live through brutal electrical, wind, and ice storms.
- Use either organic material or sunshine for energy
- Go dormant when conditions degrade, yet survive.
- Grow rapidly at any latitude, longitude or altitude.
- Grow faster than any other plant on Earth.
- Maximize productivity per unit of space.
- Thrive using no fossil resources.

All living things evolved from algae so it should be no surprise that algae contain all the essential nutrients for life and vitality. A few grams of algae a day act like a nutritional supplement providing essential nutrients, vitamins, minerals, and antioxidants.

Microalgae exist as particles of various sizes, many around 5 μ (microns) in diameter. (The period at the end of this sentence is about 100 microns.) While individual alga is not visible, algae communities appear first as a cloud and then as tiny specks that are cell clusters.

Algae provide a low-fat, low-calorie, nearly cholesterol-free source of protein. Some algae such as spirulina contain up to 70% protein by dry weight which is double the protein of food grains and roughly twice the protein of meat. Unlike meat, most algae varieties provide the full complement of nine essential amino acids. Fat content is only 5-10%, a fraction of other protein sources. A chicken egg contains about 300 mg of cholesterol and 80 calories while providing the same protein as a tablespoon of the algae spirulina, which carries 1.3 mg of cholesterol and 36 calories.

Each day, algae create 70% of the world's oxygen, more than all the forests and fields combined. Algae synthesize roughly 0.8×10^{11} tons of organic matter daily, constituting about 40% of the total organic matter grown on our planet. Algae provide a full spectrum of protein, nutrients, vitamins, and minerals because trillions of hungry consumers depend on algae to sustain their vitality. Over 100 times more creatures eat algae or algae feeders than any other food. Algae deliver protein energy and other nutrients to the smallest krill as well as to the largest animal; the great blue whale.

Microalgae

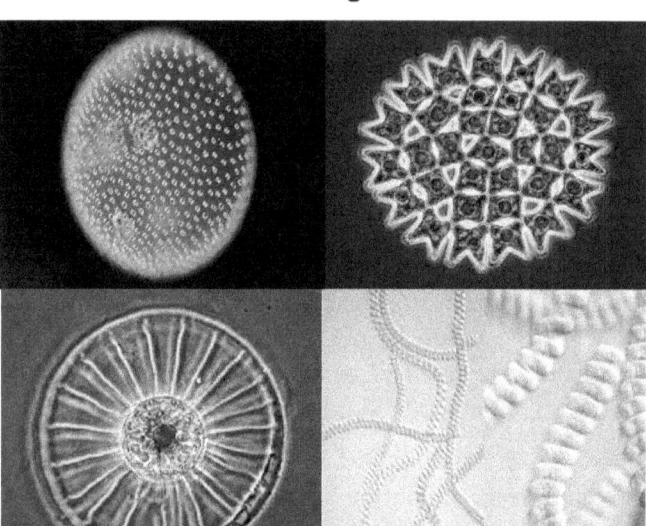

Algae grow all over the Earth in marine or fresh water habitats and on land when moisture is available. Unlike land plants that die without water, algae simply go dormant, wait for moisture and then begin their rapid development. Algae grow to the limit of the nutrient supply and then pause or go dormant until conditions improve.

Since algae form the bottom of the food chain, everything around acts as a predator. Algae's strategy to predation is brilliant – grow faster than consumers can eat. A single alga cell may produce one million offspring in a day. At night, algae take a well-deserved rest in a phase called respiration.

About 10% of algae species are macroalgae, (seaweed) such as kelp that can grow to 180 feet. Most are microalgae that occur in every color, shape, and small size imaginable. Each of the estimated 10 million algae species provides its unique nutrient profile as well as the many compounds stored in its biomass.

Macroalgae

Away from the oceans, most algae grow in, on or among the roots of land plants. A handful of local dirt may hold 100 algae species and several billion-alga cells. Land plants need algae to break down chemical fertilizers so they are bioavailable and can be absorbed by the plant. Land plants and algae work symbiotically as algae supplies nutrients and plants provide a protected area in which to grow. Algae support symbiotic relationships with mosses, fungi, yeasts, lichen, corals, and sponges. Algae attract the other microflora that populate the human stomach and break down foods so they can be absorbed.

Did algae make us human?

Scientists agree that our larger brains differentiate the human species from all others. Our larger brains give us the power of reason, planning and acting in cooperative groups. The fossil record of our human ancestors in Eastern Africa shows that after six million years of no brain size change, between two and 1.5 million years ago, the humanoid brain enlarged 300%. After brain enlargement, encephalation, our ancestors required another million years before they developed weapons for hunting or fires for cooking.

Textbooks suggest that our ancestors moved up the food chain to acquire meat, which enabled brain development. This popular macho scenario is unlikely because at 2 million years ago, our humanoid ancestors were the size of Lucy, only 3.5 feet tall. Compared to the predators that were twice as big as they are today, our ancestors were much smaller, slower, and weaker and lacked a keen sense of smell, sight or hearing. Had they tried to hunt game, they would have become one with the food chain.

Some scientists agree our ancestors could not been hunters but were scavengers. Scavenging would have been a risky and energetically costly method of acquiring food because humans lacked the speed, scent, strength, and sight capabilities of scavengers. Dead meat rots quickly and attracts parasites, worms, maggots, insects, viruses, and bacteria that would have been deadly for early humanoids. Hauling scavenged meat or bones back to camp would have been problematic with stealth predators continually on the prowl.

Instead of going up the food chain to acquire meat, our ancestors may have gone down the food chain to algae. A tribe of early hominoids may have ingested algae (probably spirulina) incidentally in the green sweet drinking water of the soda lakes in the Rift Valley of Ethiopia. They may have continued to drink the water for the sweet taste and because algae created a feeling of satiation and improved digestion. Incidental ingestion would not have provided sufficient dietary protein but may have provided essential vitamins, minerals, and antioxidants to support brain, heart, and eye development.

As their brains enlarged, early *Homo* may have exploited the lacustrine ecosystem for algae and algae feeders loaded with algal protein and nutrients. Algal nutrients were available locally, easy to harvest year round and ready to eat. Algae may have served as the first sweet convenience food and could have provided a full set of critical amino acids, essential fatty acids, vitamins, and minerals to support brain and body development.

Access to a convenient, tasty, ready-to-eat food source may have reduced the time, energetic and predation costs of moving camp. Algae consumption may have helped early hominoids avoid health drag from the micronutrient and vitamin deficiencies common in developing countries today including malnutrition, anemia, xerophthalmia, goiter, cretinism, dwarfism, and scurvy. Healthier minds and bodies supported by algae that acted as a natural food supplement may have enhanced sexual activity which would have improved survival advantage.

Algae food

For thousands of years, coastal people have harvested seaweed growing in natural stands for food, fodder, cooking fires, fertilizers, vitamins, minerals, and medicines. Indigenous people used seaweed for treatment of bruises, burns, cuts, skin irritations, and indigestion.

Dried seaweed has been used for centuries for trade because it is nutritious; easily transported long distances and has a long shelf life. Dried algae provided the first portable convenience food and probably served as wampum in trade, along with white shell beads. Archaeological evidence shows early Neanderthals around the Mediterranean ate algae along with shellfish.

Most societies around the Pacific Rim have thousands of years of experience in using algae (seaweeds or sea vegetables) as food – fresh or dried. When dried algae is put in water, the original color, which may be purple, red, brown, yellow or green, reappears. Some seaweed is not eaten directly as food but is used to flavor salads, soups, and stews or as a condiment for meats or rice.

Algae contain a wide spectrum of prophylactic and therapeutic factors that include vitamins, minerals, amino acids, essential fatty acids, the super anti-oxidants such as β-carotene, vitamins A, B, B-complex, D, E and K, and a number of unexplored bioactive compounds. Since these many constituents stimulate numerous metabolic pathways, algae has been demonstrated to promote antioxidant, anti-bacterial, antiviral, anticancer, anti-inflammatory, anti-allergic, and anti-diabetic actions, as well as vascular, mental, and intestinal health.

Unlike meat, algae provide the complete set of all nine essential amino acids. Algae are a superior protein source among plant foods, particularly the red, green and blue-green algae which offer up to 70% protein (dry weight) which exceeds soybeans (40%) and eggs (50%). Algae are also an excellent plant source of glutamic acid, an amino acid that promotes intestinal health and immune function.

Abundant Agriculture

Algae's rich set of nutrients, antioxidants, enzymes, and extracts, boost the immune system and enhance the body's ability to grow new blood cells. Algae are rich in phytonutrients and functional nutrients that activate digestive and immune systems. Algae compounds accelerate production of the humoral system (antibodies and cytokines), allowing it to better protect against invading germs. Algal components also activate the cellular immune system including T-cells, macrophages, B-cells, and anti-cancer natural killer cells.

Algae are not a primary source of calories, rather a source of bioactive agents that facilitate efficient and healthful metabolic processes. The fibrous components of algae add bulk to the digestive tract reducing hunger, transit time and intestinal pathologies. Specific algae polysaccharides have demonstrated antiatherosclerotic functions reducing blood LDL cholesterol concentrations and cardiovascular disease risk. The total fiber content of algae (~6 g/100g) is greater than that of fruits and vegetables promoted today for fiber content: prunes (2.4 g), cabbage (2.9 g), apples (2.0 g), and brown rice (3.8 g).

Algae absorb a wealth of mineral elements that concentrate as about one third of its dry biomass. The mineral macronutrients include sodium, calcium, magnesium, potassium, chlorine, sulfur and phosphorus while the micronutrients include iodine, iron, zinc, copper, selenium, molybdenum, fluoride, manganese, boron, nickel, and cobalt. On average, 1 tablespoon dried algae provides the same amount of:

- Calcium as ½ cup milk, 1½ cup soybeans, 8 carrots, or 22 tomatoes.
- Magnesium as 2½ cups milk, ½ cup soybeans, 9 carrots, or 6 tomatoes.
- Iron as 32 cups milk, ⅓ cup soybeans, 11 carrots, or 5 tomatoes.

Algae have been providing these nutritional and therapeutic benefits for its many hungry consumers for eons. Recent breakthroughs in biotechnology and biological medicine have enable scientists to recognize the substantial health benefits provided by algae foods. Our next challenge will be producing superb algae foods in abundance.

What is Abundant Agriculture?

Abundant agriculture – abundance

Abundance takes its name from the production inputs – plentiful resources that will not run out – sunshine, CO_2, waste, brine or ocean water. Nature is not cheap but smart in designing her production system with recycled natural resources that replenish nutrients from each growing cycle, while also regenerating the ecosystem. Nature's frugal yet effective design has demonstrated sustained production success over 3.7 billion years in recovering, recycling, and reusing waste stream inputs.

Algae use solar energy efficiently to transform water, CO_2 and nutrients into a green biomass rich in lipids, sugars, proteins, carbohydrates and other valuable organic compounds, Figure 2.2. Algae convert inorganic substances such as carbon, nitrogen, phosphorus, sulfur, iron, and trace elements into organic matter such as green, blue-green, red, brown or other color biomass. Each ton of algae sequesters nearly two tons of CO_2.

Figure 2.2. Algae Converts CO_2 to O_2

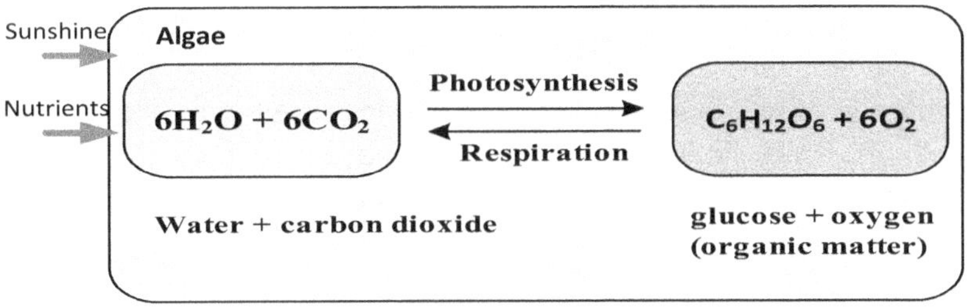

Sunshine Algae

Nutrients

$$6H_2O + 6CO_2$$

Photosynthesis →

← Respiration

$$C_6H_{12}O_6 + 6O_2$$

Water + carbon dioxide

glucose + oxygen (organic matter)

Algae form the base of the food chain in nature and supply a full set of critical micronutrients, vitamins, minerals, and antioxidants. The lipids can be squeezed from algae to make oil useful for cooking and heating and large production systems can produce clean liquid transportation fuels. The carbohydrates my produce paper, building products, textiles, biodegradable bioplastic, or be refined to fuels.

Abundance provides over 30 to 50 times more protein per unit of land than the most productive land crops including soy, corn, rice, and wheat. Abundance production uses no or minimal cropland and may grow biomass in rural wasteland or cities on any surfaces that receive sunshine. A single microfarm acre on non-crop land produces the equivalent protein of 30 to 50 cropland acres of corn or other food grain. Each pound of algae biomass has roughly double the protein available from a pound of food grain. Abundance farming for food produces a biomass that may be 60% protein and 40% a combination of lipids, (oils), carbohydrates, pigments, vitamins, minerals, antioxidants, and other valuable coproducts.

> *If one way be better than another, that you may be sure*
> *is nature's way.*   *– Aristotle*

Algae biomass production requires no or minimal fossil natural resources so abundance does not compete with industrial agriculture for resources. Modern farmers may adopt abundance to reclaim the energy and nutrients in their farm's waste stream. Farm wastes such as manure retain 60% of the original plant energy and often over 80% of the original plant nutrients. Algae assimilate residual energy and nutrients and provide an immediately bioavailable organic fertilizer back to the fields, while substantially reducing ecosystem pollution.

Abundant production systems can clean waste or brine water, providing clean water in addition to the green biomass. Coal or gas-fired power plants, cement, manufacturing, breweries, and other CO_2 polluters can flow their smoke plume through a microfarm to sequester or recycle the carbon and the black soot particulates in algae. Every ton of algae consumes nearly two tons of CO_2.

Abundance farmers use 360 microfarms and train microflora to become biofactories that produce a wide diversity of valuable products using primarily plentiful resources. Abundance closes the nutrient cycle and diminishes the consumption of fossil resources by recycling water and nutrients. Farmers practicing abundance use 360 algae microfarms that close the nutrient loop (creating a full 360) by using farm and other waste stream inputs, Figure 2.3. Microfarmers recycle the water used to grow algae, which minimizes water consumption, enables full use of the precious nutrients and eliminates run-off pollution.

Figure 2.3 Abundant Agriculture with 360 Microfarms

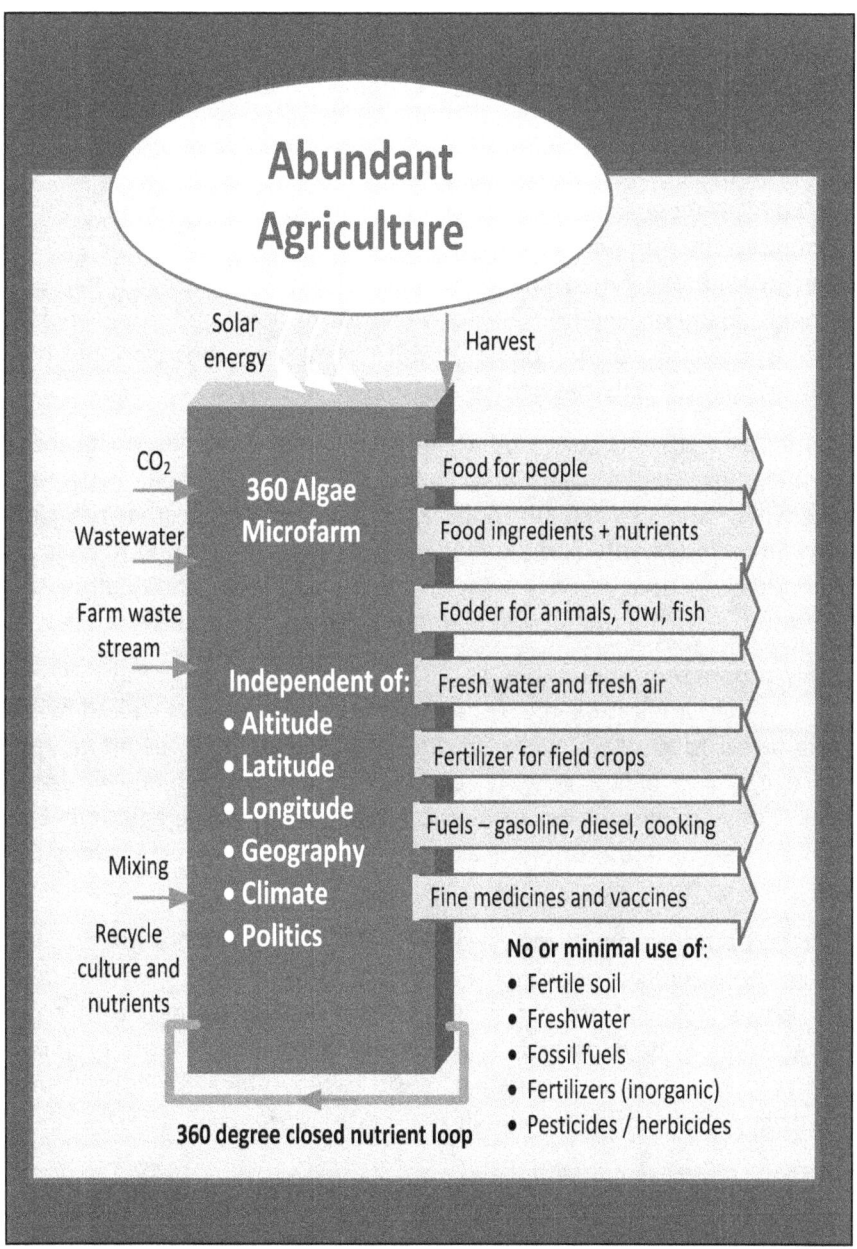

Algae propagate by a variety of sexual and asexual methods. A single alga cell may produce over one million offspring in a day. Since the organism multiplies so fast, a single farmer or university laboratory can create enough algae culture to supply 1,000 abundance farmers with start-up culture.

Food crops require substantial human time and labor over the 120-140 day growing season but algae do not because they multiply their biomass daily. Abundance farmers do not have to wait until the end of the growing season to collect their crop because they can harvest biomass daily or several times a week. More advanced applications, such as pressing out fuel oil necessitate more strenuous work but the timing of those activities can be scheduled monthly.

The human labor required by abundance is more like gardening than farming or mining. Avoiding heavy equipment, huge engines, chemical, and particulate pollutants removes most the risk for accidents and health. Some unanticipated issues are certain to arise in abundance production but most of the physical labor and health risks are removed from the food and energy production models.

Daily production operations, harvest and drying may take an hour a day, which minimizes the tedium and exhaustion associated with traditional food production. Production can stop and restart at the farmer's convenience with only a modest productivity loss. Abundance minimizes risk of physical injury because the equipment is not dangerous and monotony and fatigue are not factors. Farmers do not have to handle poisonous chemicals.

Daily production minimizes psychological and financial risk for farmers since there is low probability of crop loss. In the event of the loss of a culture, a new culture can restart in the next few days, with new production in days or weeks. After the capital cost of the growing system, farmers have modest costs when they can site their microfarm near sufficient supplies of waste nutrients.

Microfarmers use cultivated algae production systems, (CAPS) to grow algae. Microfarmers cultivate microbial communities which may be pure strains of algae but are often diverse communities of algae and the multitude of other microorganisms algae attract. Microflora communities thrive in aquatic and moist terrestrial settings and include algae, fungi, bacteria, viruses, slimes, and other tiny organisms. This diverse array of microorganisms works symbiotically to produce compounds valuable to plants, animals and humans.

Algae Production Systems, CAPS

Many CAPS designs may be viewed at www.AlgaeCompetition.com.

Algae energy

Abundance cultivates the fastest-growing plant on the planet to provide portable energy that may be used in a multitude of ways, including:

- **People** – organic protein, nutrients, and micronutrients in food.
- **Animals** – organic protein and nutrients in fodder.
- **Fowl** – natural protein and nutrients for birds.
- **Fish** – natural protein and nutrients in fish feed.
- **Land plants** – rich, full spectrum organic fertilizer.
- **Fire** – high energy algae oil for cooking and heating.
- **Cars** – lipids and carbohydrates refined to biofuels.
- **Trucks and tractors** – high energy clean, green diesel.
- **Trains, boats and ships** – high energy clean diesel.
- **Planes** – high energy, clean aviation gas and jet fuel.

Algae's food value has been known for centuries and food potential for at least 100 years. Consider the annual protein production per acre for food grains, Figure 2.4. Annual protein production per acre of algae is calculated based on a photosynthetic capacity of 15,000 pounds of protein per year. In addition to higher protein production, algae provide a superior set of vitamins and minerals than found in land plants. Algae are not a full solution for hunger and malnutrition because the biomass is low on calories.

Figure 2.4 Algae Protein Production Potential –
Pounds of protein per Acre per Year

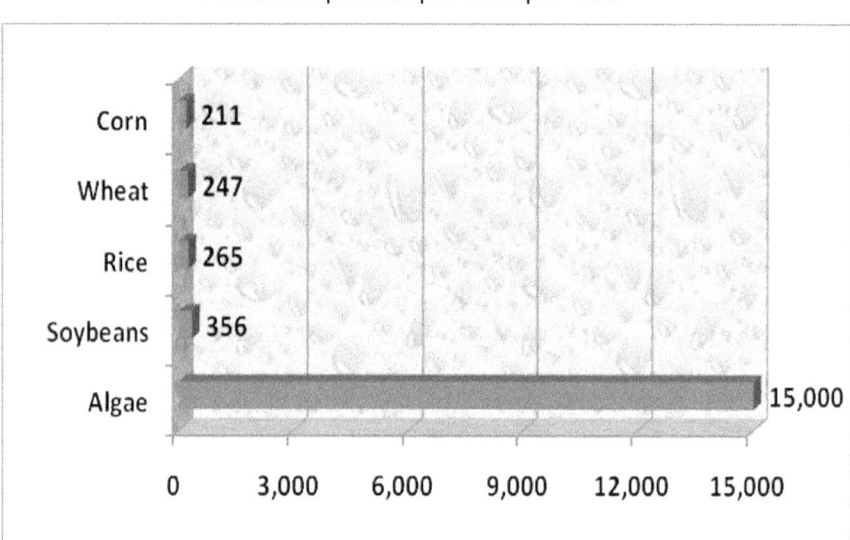

Any food, feed, textile or medicine made from land plants can be made from algae because land plants evolved from algae 500 million years ago. Algae pigments provide the pink in salmon and the striking flamingo orange in flamingos' feathers.

Algae produce the omega 3 fatty acid found in fish oil. Farmed fish that lack access to algae produce no omega 3s. Most of the fish oil sold today comes directly from algae because producers can grow omega 3s faster, easier, and cheaper in algae than harvesting fish. Fish oil from fish has two disadvantages besides cost; fishy taste and mercury accumulation. Fish oil produced from algae has a sweet organic or neutral taste and does not accumulate mercury because algae is cultivated in non-mercury environments.

Algae flours are extremely malleable in the sense that algae can substitute for wheat, corn, rice, or soy products while providing more protein and a higher nutrient density or nutralence. Algae foods may include protein-rich milk, ice cream, chocolate (with superb taste and 80% less fat), baked goods of any size, shape, or texture such as tortillas, crackers or cakes. The biomass may be made into texturized vegetable protein with added fiber. The high nutralence biomass may be extruded to make additives for meats that improve moisture retention and increase protein while lowering fat.

Processing algae can match the form of nearly any food such as peanuts, pasta, pesto or protein bars. Fortunately, years of food processing experience with terrestrial crops that have an unappealing natural taste such as soybeans make it easy to add colors, flavors, textures (fibers), and aromas.

Any product that can be made from fossil fuels can be made from algae because nature chose algae as the primary feedstock for fossil fuels. Commercial producers are excited about replacing fossil fuels with algae. However, human societies survived for many millennia without the convenience energy sources derived from fossil fuels. The most critical energy source for humans is food because we survive only a short time when deprived of the vital energy supplied by food.

Abundance and sustainability

Abundance addresses each of the sustainability issues associated with industrial agriculture, Table 2.1.

Table 2.1 Abundant Agriculture Sustainability

Health and food security
1. Nutrition and health

Algae-based foods provide substantially more protein per pound than food grains or meat, along with a far superior nutralence (nutrient profile). Algae provide a full spectrum of micronutrients, trace elements, and antioxidants not provided by most traditional foods. Algae biomass can be made into veggie burgers, chips, snacks, chocolate, and ice cream that have superior taste, 80% fewer calories and substantial nutritional benefits. Algae foods can end nutrient dilution and add vital nutrients to diets.

Algae slow the release of sugars from the blood stream, benefiting pre-diabetics. A few grams of algae create a feeling of satiation, reducing the cravings that lead to over eating. The natural vitamins and omega 3 in algae supports brain, eye, and heart growth, development, and functioning throughout life.

Algae can clean ecosystems polluted by industrial agriculture, which promises to improve health for aquatic and human life.

2. Food security

Distributed food production diffused among many producers improves the reliability of food availability and affordability. Robust algae species can usually be found locally, creating a tremendous variety of algae cultivars. Avoidance of monocultures further enhances food quality and production stability.

Algae farmers can produce around 15,000 pounds of protein per acre each year and will enhance food security substantially. Production local to population centers will reduce food cost and enhance availability. Growing food without fossil inputs will enable food supplies even when fossil resources are short, expensive or unavailable. Distributed cultivation overcomes the significant problems of politics and the food supply chain.

3. Climate chaos

Abundance practiced in closed or semi closed CAPS produces energy and nutrient rich biomass independent of latitude, longitude, altitude, geography, climate or politics. Many algae species produce well in cold climates. Algae grow in all the oceans, under both ice caps, in wetlands, mountains, plains, valleys, and deserts of the world. Microfarms require sunshine or possibly grow lights. Microfarms in high latitudes may have highest production in a greenhouse.

4. Risk – economic, physical and health

Abundance reduces food production risk as farmers may harvest half of their rich biomass daily or every several days. If weed algae or predators invade their CAPS, microfarmers clean the system and start a new culture immediately. If production is interrupted due to a religious holiday, festival, travel or illness, new production can begin quickly when the farmer returns. Production slows or stops without sunshine but begins again immediately when the sun reappears. Microfarmers in cloudy areas may use grow lights or LEDs to substitute for solar energy. Some microfarmers cultivate heterotrophic algae that grow in dark containers using plant sugars as the energy source.

Microfarming requires about the same physical labor as gardening, which enables the elderly, women and disabled people to produce their own food. Growers may tend to their culture for hour or two each day, so they avoid monotony and fatigue. No heavy equipment is necessary and no agricultural poisons jeopardize grower health or the health of neighboring communities.

Economics and overconsumption

5. Costs

After the modest cost for constructing the CAPS, production inputs are plentiful and affordable: sunshine, CO_2, and wastewater. If waste CO_2 is not available locally, growers may burn organic wastes in a closed kiln such as a pyrolysis unit and flue the CO_2 to the CAPS. Operational costs for microfarms are low, especially when renewable energy is used to mix the culture and for harvest.

6. Fossil resource consumption

Abundance produces food and other form of energy using no or minimal fossil resources. Production in closed systems may occur on land unsuitable for crop production such as wasteland, hills or deserts. In cities, CAPS can be placed on rooftops, balconies, the sides of buildings or in buildings, such as vertical farms.

CAPS may be open or covered ponds, troughs in the soil, or closed or partially closed systems that enable solar energy or grow lights to provide the energy for photosynthesis. CAPS may be built in nearly any size laterally or structured as vertical farms. No or minimal fresh water is needed because algae thrive in grey, waste, brine, brackish, and ocean water.

Abundant production requires modest mechanical energy for moving and mixing the water, CO_2, and nutrients as well as for extraction and coproducts separation. Energy can come from renewable sources such as solar, wind, waves or geothermal. Agricultural, municipal or industrial waste streams as well as brine or ocean water can provide sufficient fertilizer nutrients for biomass production. Abundance eliminates the need for agricultural poisons such as herbicides, pesticides, and fungicides.

7. Ecologically positive

Production in closed or partially closed CAPS minimizes or eliminates air, soil, and water pollution. Open CAPS, such as ponds, also have a positive ecological impact. The only gas given off by biomass production is pure oxygen. Microfarmers keep the culture water moving which minimizes predators such as bacteria. The awful smell associated with algae blooms in waterways comes not from the algae but from the bacteria that attack the algae and consume the oxygen algae generate in the water.

8. Transportation

Microfarms practicing abundance and SAFE production in or near cities reduce transportation energy and cost by roughly 80%. Removing food trucks from cities reduces the carbon and black soot load in the air as well as traffic congestion. Microfarms can be placed on rooftops, empty lots, balconies on the sunny side of buildings or any other place that receives sunlight. In rural areas, CAPS may be placed nearly anywhere and need not compete for space with cropland.

Social – Access to food

9. Social equity

All people, including women, minorities, uneducated and disabled people, have access to the plentiful and cheap inputs needed for SAFE production. The light physical requirements assure that most people have the physical capacity to practice abundance.

Diffusion of abundance knowledge globally will enable those people with desire to produce sustainable and affordable food and energy for the needs of their family and their community locally.

10. Hunger and malnutrition

Distributed production of good food in the hands of those most in need in rural and urban settings can moderate poverty, hunger, and malnutrition. Algae can reclaim energy and nutrients from waste, brine or ocean water, and concentrate elements such as iodine at 1000 times ambient levels. Therefore, communities that get insufficient iodine from local food or water (because it is to dilute) can avoid nutrient deficiency by eating a few grams of algae daily. The same opportunity holds for other nutrient deficiencies such as iron, zinc, magnesium, vitamins and trace elements.

Abundance production may also serve to moderate the impacts of famine, fierce storms, floods, earthquakes and other natural calamities where people need access to good food. Setting people up to produce their own food diminishes the cost of food aid.

Sustainability scorecard

Abundant agriculture has not been practiced sufficiently to prove the economics or sustainability. Current practice shows promise for robust sustainability. The history of algae as food, feed, fertilizer, nutrients, pollution solutions, and medicines have demonstrated high value for nutrition and health, fossil resource consumption, soil degradation, and pollution as well as resolving nutrient deficiencies and malnutrition.

The sustainability scorecard in Table 2.2 represents what I believe are obtainable goals rather than current practice. We will have to collaborate locally, regionally, and globally to achieve these goals.

Table 2.2 Abundance Food Sustainability Scorecard

10 = Highly sustainable

Health and food security	Grade /10
1. Nutrition and health	9
2. Food security	8
3. Climate chaos	8
4. Risk – economic, physical and health	9
Economics and overconsumption	
5. Costs	7
6. Fossil resource consumption	8
7. Ecologically positive	10
8. Transportation	9
Social – Access to food	
9. Social equity	7
10. Hunger and malnutrition	7
Total	**82/100**

Abundant agriculture does not receive perfect scores because there are inevitable trade-offs in food production. Food production costs will rise systemically for all forms of food production but should be substantially less for abundance growers that avoid most the costs associated with fossil inputs. Abundance can reduce social inequity and hunger but offers no mechanism to slow population growth.

Chapter 3. Why Abundance only Now?

God Almighty first planted a garden. And indeed, it is the purest of human pleasures.

— Sir Francis Bacon

The first question most people ask about abundance food is: "Why wasn't abundance invented before?" Consumer behavior, the science of why people make certain choices, provides an answer. Farmers have been influenced by agribusinesses promising that synthetic chemical inputs trump nature. Unfortunately, no one bothered to support the research to examine the efficacy of synthetic chemicals compared with natural processes such as algae.

Abundance represents a "natural process" invented by nature and cannot be patented or listed as intellectual property. Agribusinesses that supply inputs to farmers hire legions of technologists and scientists to develop synthetic compounds that can be patented and sold for extremely high premiums. Agribusiness advertising promotes this intellectual property and creates a belief in better living and farming through chemistry. Monsanto's Round-Up™ is among the most common words in modern farming.

Consumers are addicted to highly processed foods with high sugar, fat, and salt but few nutrients. Farmers are addicted to increasingly expensive genetically engineered seeds, synthetic fertilizers, herbicides, pesticides, and fungicides. The sad irony is that the synthetic chemicals kill the beneficial microbes that nature put in the field to feed and nurture plants. Imagine, paying premiums to kill the organisms that work symbiotically with plants to provide nutrients and plant hormones for their vitality and defense.

Modern farmers have bought into chemical fertilizers because they are cheap and easy to apply. Industrial agriculture systemically extracts macro and micronutrients as well as organics from field soils. Each year, most farmers replace only the big three NPK macronutrients (nitrogen, phosphorus, and potassium). With every crop, micronutrients diminish along with soil organics, which creates nutrient dilution and the loss of color, taste, and texture in produce.

Modern farmers buy large, heavy tractors that cultivate quickly but compact soil, which diminishes root growth and accelerates erosion. Farmers buy tons of chemical fertilizers, pesticides, herbicides, and fungicides. Plants are developing resistance to chemical fertilizers so farmers must apply more. Pest and weeds are developing resistance to chemical poisons which means farmers must use more or change poisons. Plants often absorb less than 1% of the agricultural poisons applied to fields, which creates enormous waste and cost. The residual fertilizers and poisons flow into wetlands, streams and groundwater where they disrupt, degrade, and destroy local ecology.

Abundance is antithetical to agribusinesses because natural, bioavailable algae fertilizers and other plant inputs are natural and cheap. Nature was engineering marvelous products that meet plant needs long before Monsanto entered the business. Algae and the symbiotic microbes they attract create or provide the compounds that enable plants to naturally synthesize many of the advanced compounds they need to grow and to fight disease and pest vectors. Unfortunately, US government farm policy has chosen to support R&D on industrial food production rather than natural processes.

Why Abundance only Now?

Farm policy

Farm policies, government sponsored research to Land Grant universities, extension service agents, subsidies and food support for the hungry, play a major role in food production. Farm policies have been dictated by the same large agribusinesses that have addicted farmers to their branded synthetic chemistry. Wealthy farmers and agribusiness like ADM, Monsanto and Cargill make enormous political donations to both parties in order to shape policy in commercial agriculture to benefit their interests. Consequently, over 99% of federal grants and R&D go to industrial agriculture. Unsurprisingly, extension agents who are in place to help farmers and gardeners are trained in industrial production. Less than 1% of federal funding goes to organic production, R&D or training. Government support in the US for natural processes such as abundance over the last 30 years has been near zero.

India and China support natural processes R&D in food production because their leaders realize that fossil resources are finite, increasing in price and will eventually run out. Both countries have terminated their biofuels efforts with food crops for the obvious reason that food-based biofuels drive up the cost of food and the inputs to produce food. China recently put a 135% tariff on their phosphorus fertilizer to insure sufficient supplies for domestic farmers. India's scientists have performed some excellent R&D with natural biofertilizers, especially focused on cyanobacteria that fix nitrogen and reduce the need for chemical fertilizers.

As modern farm policies evolved in the 1960s to 1980s, the food supply sustainability issues were not well articulated. Consumers and political leaders preferred celebrating their brilliance in designing the Green Agricultural Revolution and the cheap food it provided. Leaders and policy makers preferred to ignore critical issues with genetically improved seeds such as the need for additional cultivation, two to three times more irrigation, triple the need for fertilizer and 20 times the need for agricultural chemicals and poisons.

Few people were aware of nutrition and health issues, food security, fossil resource depletion or global warming before the 1980s. The few voices that challenged the fossil foods path such as Michael Pollan and Robert Henrikson were drowned in the wind of political rhetoric. Even today, only one third of Americans believe the scientific consensus that global warming is caused by humans. However, neither politicians nor consumers can deny that humans caused severe fossil resource depletion with our cheap food policies.

Biofuels

In the 1990s, the Clinton administration made a critical political mistake and stopped R&D on algae for food or biofuels and shifted those resources to corn ethanol for biofuel. The decision by the EPA to fund a corn ethanol industry may have been the most costly decision in American history because it accelerates natural resource depletion. When the US runs out of resources to produce food, who will sell us food? Where will the government find the money to buy food for hungry Americans?

The farm lobby remains so strong that corn ethanol subsidies continue at around $20 billion a year, even though ethanol consumes more fossil energy than it returns. Huge subsidies flow primarily to large agribusiness and land owners, not to family farms. Subsidies continue in spite of clear scientific proof that corn ethanol is an expensive, wasteful proposition that not only massively depletes our natural resources but creates billions of dollars in degraded and damaged ecosystems.

The 44 million acres of corn grown for ethanol in 2010 could and should be replaced by less than 2 million acres of algae biofuel production while improving air and water quality.

Biowar I: Why Battles over Food and Fuel Lead to World Hunger (Edwards, 2007) traces the money path, primarily to one company, ADM, that initiated the biofuel industry with millions in political donations to both parties. Today ADM receives billions each year in biofuel subsidies.

A biowar occurs when a country burns food, typically as a horrific act of war on another country. In the case of Biowar I, the US became the first country to burn its own food. Biowar I ignited when the Bush

Administration announced the Energy Policy Initiative in 2005, which increased biofuel subsidies and mandates, Figure 4.1. The unintended consequence of producing large amounts of corn ethanol on US and world food markets was predictably higher food prices. In the eyes of the UN, World Bank and most foreign countries, the US ethanol policy contributed substantially to the terrible 2008 food riots in 40 countries.

Figure 4.1 Corn Burned for Ethanol and Food Stamps

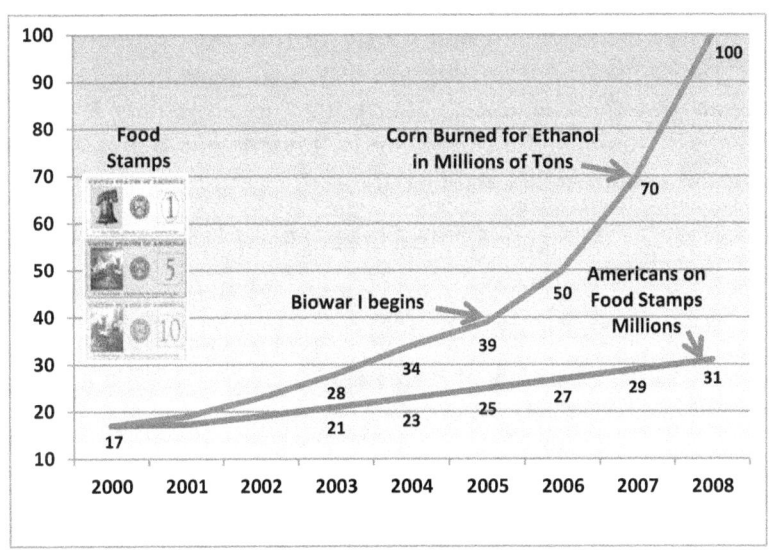

How could people in hungry countries not blame the US for food shortages and price increases when prior to the ethanol program, America provided half the world's food grains and 70% of the world's corn imports?

How does a country with over 60 million people receiving food support because they are hungry justify a policy of burning its citizen's food for a weak fuel additive? Over 30 million Americans are on food stamps and must abhor the concept of burning food because they know their $1 a meal buys less food. Biowar I can end by a simple government policy that states: "Subsidies should flow to sustainable, non-pollutive food and energy production systems."

In 2009, the US became a net importer of food. A college sophomore could make the case that the US biofuel policy is wasteful and foolish.

Cheap food

Abundance makes little sense when the cost consumers pay for industrial foods are apparently so cheap. Appearances can be deceiving. Fossil foods appear to be inexpensive because government subsidies and poor math make food appear cheap to consumers.

Governments lavishly subsidize industrial farming, big agribusinesses, big oil, water management and the fossil resources on which food production depends. These subsidies in the US reduce the real food cost by about a third, Figure 3.2. Currently, these subsidies are being financed with our children's money in government bonds held by countries like Saudi Arabia and China.

Figure 4.2 Real Cost of Food

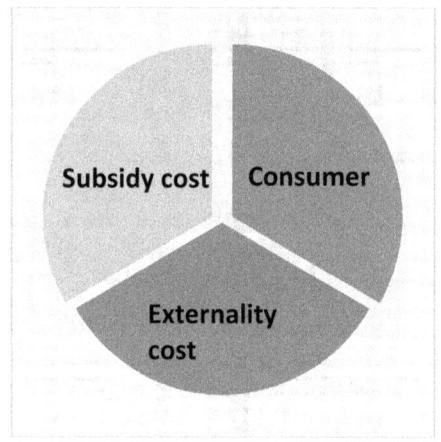

American corn subsidies decimated Haitian farmers because they could not grow food as cheap US food dumped on the country as "food aid." The US corn subsidies also have displaced over a 1.5 million poor Mexican farmers. Farmers were forced to leave their land because they could not compete with subsidized U.S. corn. Many of these farmers added their feet to the flow of illegal immigrants to the U.S. from Mexico. Canada, Mexico and other countries have outstanding lawsuits against US subsidies with the World Trade

Why Abundance only Now?

Organization because these subsidies substantially depress the real price of food grains.

British International Development Secretary Douglas Alexander said "It's unacceptable that rich countries still subsidize farming at $1 billion a day, costing poor farmers in developing countries $100 billion a year in lost income." The US subsidies hide the true cost of food, which benefits consumers in the near term but bloats the financial deficit in the intermediate and long term.

A group of more than 400 agricultural experts, known as the International Assessment of Agricultural Knowledge, Science and Technology for Development concluded through its global and regional studies report that governments and industries need to discontinue environmentally damaging farming methods. At their 2008 meeting in Johannesburg South Africa, the group recommended "ending subsidies that encourage unsustainable practices." Political leaders in the US should listen to world opinion because US subsidies will destroy our ability to grow our own food in the near future.

Another third of the true food cost comes from externalities such as resource depletion and environmental degradation, for which the food supply system fails to account. Farmers pay for neither resource depletion nor ecosystem pollution. Environmental degradation alone creates about $45 billion a year in damage. No metrics are currently available for resource depletion. Neither farmers nor consumers pay a nickel for these costs, so they are shifted to US society.

When our high school and college students take to the streets and demand change, policy leaders will realize we are over-consuming and wasting our children's non-replaceable natural resources required for industrial food production.

A full lifecycle accounting would show fossil foods are substantially more expensive than abundance production. Life cycle accounting includes the economic impact of degrading air, water, and soil, destroying our fisheries, creating dead zones as well as cost to human and animal quality of life and health. The current generation benefits from resource depletion. We turn a blind eye to resource loss by not accounting for depletion in the price of our food. The next generation will not enjoy the same luxury.

When our children discover industrial agriculture lacks the natural resources to produce food, they will ask the government for increased subsidies. Unfortunately, the government will be out of funds. What country would be willing to make loans that add to the immense US debt? The US is already a debtor nation; we just fail to act as one.

When our children discover their fields are worn out, fresh water is unavailable, fuel costs are out of reach, fertilizer mines are exhausted and agricultural chemicals have ruined their waterways – will they agree that our fossil foods were cheap?

We may have developed a less expensive, non-fossil food production system had people not had distaste for one word – algae.

Consumer behavior and algae

Why do most people have an immediate aversion to algae? The answer is false attribution. Would you have picked up this book if it were titled: *Algae Agriculture*?

When asked to describe algae, people's top of mind typically elicits several words with strong negative connotations: "slimy, smelly, scummy, and yucky." If putrefied raw meat were presented as steak, people would naturally dislike steak.

People falsely attribute the smell in ponds to algae because it certainly looks like algae. Actually, the odor comes not from algae but from the bacteria that attack and eat the algae. The bacteria consume all the oxygen algae added to the water, causing entrophication. When the aquatic organisms are deprived of oxygen, they die and begin rotting, which adds smell and slime to ponds. Healthy algae give off lots of oxygen and smell similar to walking through a redwood forest – without the redwood trees, of course.

Weed algae in ponds grow in diverse communities of many microorganisms and are different from the algae we cultivate in abundance microfarms. Gardeners know they must remove the weeds from the garden or the weeds will take over the garden and consume all the nutrients. Abundance growers control weed algae in order to enable healthy production of the target species.

The yucky factor seldom occurs with edible algae. At a recent Rotary meeting at a Kobe Steakhouse, the Rotarians were served three forms of algae at the luncheon: algae soup broth for taste and thickening, sea vegetable in the soup for texture, color, and visual appeal and separately a cold seaweed salad for color, taste, and texture. The Rotarians were asked after lunch if anyone liked algae to eat, which elicited the yuck factor, as all were extremely negative. They were quite surprised yet delighted with the colors, taste and texture provided by the sea vegetables they had been served. The Kobe manager brought out a large variety plate of sushi made with algae. The sushi plate with various forms of algae was consumed in one pass.

Nori represents only one of more than 1,000 sea vegetable and has a world market value of over $3 billion a year. Nori serves as a luxury food and is often wrapped around the rice ball of sushi or rice with a slice of raw fish on the top. Toasting or baking brings out Nori's rich flavor and flakes are often sprinkled over boiled rice or noodles. Nori may be incorporated in soy sauce and boiled down to make the rich sauce. Nori is also used as an ingredient for jam and wine. Chinese cooks use Nori in soups and for seasoning fried foods. Many other marine and freshwater algae foods await commercialization.

Abundance critiques

If abundance methods were easy, they would have been developed and diffused decades ago. We are just beginning the development of the practice, abundant agriculture 1.0. Initial critiques of abundance include the following.

Table 3.1 Abundance Limitations

1. Farmers will never give up dirt farming and adopt abundance.

Farmers do not have to abandon their fields or favorite crops. Farmers can regenerate their fields and improve crop yields. Growers want to leave their fields better than they found them.

2. We will not run out of fossil resources.

Really? The question isn't whether we will run out, only when. In many developing countries, fossil resources, especially productive seeds, water, and fertilizer, have already become too expensive.

3. Algae production is still 10 years away.

No, firms have been producing algae profitably for 30 years. We conducted the only algae industry surveys done to date, (for the Algal Biomass Organization and for Algae World) and those results show over half of producers believe successful algae production is operational now or will be within three years. Over half of respondents believe that the US to replace ethanol with algae in under 10 years. The algae producers profiled in Chapter 5 are not unique. Over 150 companies are producing algae successfully today. Once we get CAPS designed for gardeners, abundance growth will become exponential. Our goal is to motivate one million Green Masterminds (algae producers) within 10 years and 10 million within 15 years.

4. Algae production is too hard.

Yes, abundance has been too difficult for nonscientists but that is changing. I have failed three times over the last 20 years in constructing a backyard CAPS. All we need is one success – and a person or team willing to share the techniques as open source. We are engaging Green Masterminds at www.AlgaeCompetition.com to share their innovative designs and production breakthroughs. Please join the fun and become a full-fledged Green Mastermind.

We have successfully prototyped smartcultures production near fields. Crop yields and cost reductions exceeded our expectations. However, we need to make the CAPS less expensive, more portable and easier to operate and maintain. These upgrades should occur in 2011. Diffusion of innovation before the CAPS operate reliably and efficiently would create the same false promise that has doomed many prior world food initiatives.

5. Algae production is too expensive.

No, fossil agriculture is too expensive because consumers pay only one third the cost of food. One third is hidden in subsidies and another third, ecological cost, is unaccounted for and ignored. A lifecycle analysis of fossil foods that accounts for all costs including extraction, depletion, waste, erosion, ecological pollution, and all the impacts to human and animal health, will show that industrial agriculture is more expensive than abundance. The key political question will be how long people are willing to ignore the substantial subsidy and ecological costs imposed by industrial agricultural production.

6. Isn't abundance simply organic food production?

No, organic growers avoid, to the degree possible fossil inputs such as GE seeds, chemical fertilizer and agricultural poisons. However, organics often requires more fertile soil, fresh water and fuel than industrial agriculture. Both organic and industrial growers can adopt abundance methods to improve yields, reduce costs and operate more sustainability.

7. There is no such thing as a free lunch.

Abundance production is not free in an economic sense because all inputs have some cost in terms of transportation, labor or capital. Recovering, recycling, and reusing farm waste stream nutrients closes the nutrient cycle and frees farmers from continual extraction from their fields and the need to continually purchase more fossil fertilizers, and chemicals. While abundance may not be free, the practice offers a very cheap lunch compared with the escalating costs of modern fossil agriculture.

8. People will refuse to eat algae.

True, some people will continue to eat meat and potatoes, assuming they can still afford the price of meat. Some may have cultural values that forbid ingesting microorganisms – other than the algae, fungi, bacteria, molds, mildews, yeasts, and viruses that ride naturally on everything we eat. These organisms can also be found in our gut where they work symbiotically to provide bioavailable nutrients from the food we ingest.

People who want to abstain from eating algae directly may use abundance to grow algae for aquaculture, hydroponics or aquaponics. Others may grow algae to feed their birds, dairy or meat animals. Farmers may grow algae in the smartcultures model to improve yields and quality of their field crops.

New food processing technologies will transform empty calorie snack foods into tasty, low-fat, high-nutralence health foods. A food renaissance will transform convenience foods into foods that build strong bodies and minds. Algae flour and oils will enable people to have their chocolate cake with ice cream and whipped cream – and eat it too; without fat guilt. Superb algae-based gourmet foods will become the rage in upscale restaurants. Some algae foods already on fine restaurant menus.

9. What if algae run amok, e.g. the movie *Solyent Green*?

Algae are already plentiful in the natural environment. Should some algae get loose; the culture will grow to the limits of the in situ nutrients and then stop. Algae grow only to the limit of available nutrients and then stop. Of course, we can also kill algae with chlorine, as we do in pools.

A remake of *Solyent Green* would set the algae industry back two decades. *Solyent Green* posited in 1968 that our world would become over-populated, polluted and overheated from trapped gasses, which limited our ability to grow food. The movie premise was predictive of exactly what did happen. The only possible food solution was growing algae in the massive human waste stream. The producers made the mistake of including dead humans in the waste stream, which created the fear of cannibalism.

10. If abundance were this easy, it would've already been done.

No, prior to recent biotechnology breakthroughs, abundance was impractical. Recent innovations in biophysics, biochemistry, bioengineering and a host of other disciplines have converged to make abundance possible now. Of course, we will have to apply new technologies to make production relatively easy and reliable.

11. What if algae carry a disease?

There is a legitimate concern that algae, just like any other food, may carry salmonella, e-coli or another pathogen. Continual monitoring, possibly remotely, can identify and stop cultures that may be problematic.

12. Aren't waste streams dirty?

Yes and the pathogens can be killed using natural tools such as solar heaters. These technologies have been used successfully for 50 years throughout the food supply chain.

13. Aweehhh!

Algae are just a bunch of cells. Yes, just like a carrot but with 10 times more beta-carotene per ounce. Carrot cells are differentiated while alga cells are not. Macroalgae and sea vegetables are made of many independent cells but organized as pseudo-stems and leaves similar to land plants. Algae can be made into any form, texture or color desirable, including nutritious chips, flour, snack foods or woven into synthetic meat.

14. Abundance does not address population management.

True, and without some form of population management, many are doomed to poverty and hunger. Abundance addresses large families indirectly because families in developing countries often have extra children in order to assure labor for the farm. Abundance removes the need for extra children as a form of supplying farm labor because abundance enables food production with light work that available to almost anyone.

15. Will abundance take jobs away from farmers?

No, abundance engages farmers in a new set of actions that makes farming more profitable. Many aspiring farmers can practice abundance farming because it requires minimal space and time. Abundance will create a new industry with millions of green jobs.

16. There are no businesses to support abundance.

True today, but consider the many phenomenal entrepreneurial opportunities in this new industry.

17. How can we get started now?

Use social media to explore your ideas. Create teams that design environmental landscapes, microfarms, algae foods and menus. Post your ideas at www.AlgaeCompetition.com. Use the site to extend and refine your concepts and find other like-minded social entrepreneurs.

18. The farm lobby is too strong with its huge political lobbies and guaranteed subsidies.

The farm lobby shames the U.S. Congress and all Americans who pay enormous amounts of public money, 90% of which goes to large agribusinesses and a few wealthy landowners – not hard working farmers. The farm lobby may embrace abundance in order to answer to Congress for the health of our children, the overconsumption and loss of our precious natural resources. and the pollution of our ecosystems.

19. No highly productive, low-cost, easy maintenance CAPS are on the market today.

True, which means we must design, develop, demonstrate and diffuse CAPS that anyone can use. Engineers and farmers excel at building appliances and handy tools such as refrigerators, cars, boats, and planes that all people can operate, even though many are clueless about how the tools work. CAPS need to be designed so people without language skills can learn production quickly.

20. How could we subsidize farmers to help transition to abundance?

America needs is an agriculture or homeland security secretary who has courage to make a smart policy statement:

> *Food and energy subsidies should benefit the production of sustainable and ecologically positive farming practices.*

Congress could then shift the $20 billion a year in subsidies and costs currently wasted on corn ethanol to multiple forms of truly sustainable energy.

21. You will never be able to train people to grow algae.

We do not intend to train just anybody. We plan to train Green Masterminds who share abundance passion and are willing to conduct peer training with others. We need first to train farmers and gardeners with deep knowledge about growing plants. We will need abundance demonstration sites in every community so there will be many grower opportunities.

We also need to engage scientists, technologists and students with biotechnology knowledge. We need green masterminds with a passion for improving food quality and security for their family and community. We need educators who codify the training materials and convey abundance methods to people without reading or language skills. We need social, political, business, and religious leaders who develop vision, values, and action plans to move abundance forward. We need a green environmental groups and NGOs to promote the abundance environmental and health benefits. We need media people who clearly articulate both the value proposition and the urgency for action. We need University professors and scientists to conduct the research that critically examines each of the value propositions for abundance. Most of all, we need our children to engage in this new food production model because their survival may depend on abundance.

Chapter 4. Farmer and Gardener Engagement

God Almighty first planted a garden. And indeed, it is the purest of human pleasures.

— Sir Francis Bacon

Farmers and gardeners are smart and will rapidly adopt abundance when the production models are sufficiently developed and field-tested. Abundance needs more R3D; research, development, demonstration, and diffusion. Farm decisions are driven by production expectations, costs, and market prices. Abundance can improve production while reducing costs but has no influence on commodity prices. However, an abundance farmer added over $12,000 to his bottom line because his produce in the field fertilized by algae moved up to the highest grade.

Green solar gardeners will be attracted to abundance for the larger yields of tastier and more nutritious and colorful produce. Similar to farmers, gardeners are continually experimenting with new growing methods and plant varieties. Smart gardeners want to maximize productivity while practicing predominantly organic gardening and minimizing their production costs, and use of agricultural chemicals.

Abundance farmers and gardeners producing for local farmers' markets will embrace abundance methods because they maximize local production while minimizing costs and pollution. Produce grown with abundance methods offers high-quality differentiation for abundance growers.

Our preliminary research indicates rapid farmer adoption of abundance methods will occur when farmers and gardeners know each dollar invested returns three dollars in yield improvement or cost reduction. Farmers would like to see their investment payback within two years.

A sweet spot for abundance production lies in small units for home gardeners and local farmers' markets. Home gardeners, horticulture, permaculture and hydroponic growers will transform the way crops are grown with substantial diversity and independent thinking.

Another gold mine for abundance production lies in dairy and animal farmers who can transform their substantial waste management costs into a profit center. Farm manure contains roughly 60% of the original energy that was in the crop before it went to the animal. Even more astounding, manure retains 80% or more of the original plant nutrients.

In the near future, phosphorus, (P) fertilizer will become very expensive and eventually unavailable. P is critical for all living cells but the primary body utilization occurs during adolescence with bone, teeth, and hair growth and development. Consequently, animal poop retains 95% of the P from the feed crops.

Abundant agriculture is technology neutral in the sense that superior production will require each farmer's thoughtful application of farming methods that are appropriate locally for the weather, crop, input availability, and cost, as well as markets. While abundance enables farmers to transition to predominantly organic methods with 360 microfarms, some may augment production with selected tools and techniques used in modern agriculture. Adoption of innovation research shows incremental changes occur more successfully than a radical change like a swift move from industrial to organic farming.

Abundance provides substantial value for farmers by avoiding fossil inputs, increasing productivity and reducing costs. Abundance farmers using smartcultures, grow and flow algae to their fields, and can improve soils and reduce water, energy and fertilizer waste while decreasing soil erosion and air, water and soil pollution, Table 4.1.

Table 4.1 Abundance Farmer Benefits

Benefit	Crop inputs and costs
Fossil inputs	Eliminate or minimize scarce and expensive fossil inputs including fertile soil, fresh water, fossil fuels, fertilizers and fossil agricultural chemicals.
Abundant inputs	Produce food, coproducts and other forms of energy using solar energy, CO_2, and wastewater.
Improve yields	Enhance yields of protein, lipids, carbohydrates, energy and other target compounds 20 to 50 times per unit of land each year.
Crop diversity and nutrition	Expand crop diversity providing better nutrition, micronutrients, vitamins, and minerals.
Decrease production risk	Harvest biomass daily to remove or diminish production risk.
Expand geography	Enable food and energy production anywhere on Earth with sunshine.
Grow climate independent crops	Produce effectively despite prolonged heat, droughts, more extreme storms, salt invasion, rising oceans, wild fires, and pest infestations.
Enable the poor to produce food	All people can produce food when the inputs are free or surplus, assuming they have access to growing systems and sufficient training.
Nutrient recovery and delivery	
Waste stream	Transform a cost, getting rid of agricultural wastes, to a profit center where solar energy and algae are used in to recover nutrients.

Bioavailable delivery	Deliver nutrients at the right growing cycle stage in bioavailable form that plants can use.
Smartcultures	**Boost field crop quality and productivity**
Texture and taste	Improve produce texture and taste through the immediate bioavailability of micronutrients.
Productivity	Improve crop yield, speed to maturity, size, weight and quality 30-50% by providing bioavailable nutrients.
Vitamins and minerals	Enhance the presence, quality and availability of vitamins and minerals 20-30% in produce with bioavailable nutrient and micronutrient delivery.
Digestible nutrients	Increase the presence of digestible nutrients in produce 30% with organic biofertilizers.
Smartcultures	**Upgrade soils**
Soil compaction	Reduce soil compaction and increase prosody 400% to stimulate root growth, make room for microflora and worms to enhance plant strength.
Crust	Strengthen the soil crust to add nutrients, organic material, and minimize erosion.
Soil structure	Improve topsoil structure by expanding the porosity and organic material in the soil.
Soil microbes	Use algae to attract microbial communities that act to enhance crop health and productivity.
Soil moisture retention	Improve soil moisture retention and diminish heat and drought stress.

Smartcultures	Improve agroecology
Fertilizer pollution	Reduce air, soil, and water pollution by using fewer chemical fertilizers.
Erosion	Minimize soil loss to wind and water.
Agricultural chemicals	Minimize pollution from agricultural chemicals and poisons by diminishing or eliminating them.
Bioavailable nutrients	Deliver bioavailable nutrients to the soil precisely when plants most need them.
Greenhouse gases	Reduce greenhouse gas emissions, especially CO_2, methane, and NO_x.
Tillage	Reduce the need for tillage and soil disruption.
Organic farming	Support and accelerate the transformation from industrial farming to organic farming.

Waste streams

Waste streams are the bane of many farmers existence – messy, smelly and heavy. A single cow typically creates waste that costs the farmer about $0.40 a day to move, store and eventually burn or bury. Many farmers cannot use manure on their fields because it contains too many pharmaceuticals. A dairy of 10,000 cows creates a waste management cost of $4,000 a day or $1,460,000 a year. Farmers can not only avoid the waste cost but grow green biomass containing the rich set of coproducts that the farmer may sell or use on the farm.

Nature makes sunshine free but not nutrients. Fortunately, algae have the capacity to recover energy and nutrients from the waste streams of farms, municipal waste facilities as well as power, cement, and manufacturing plants. Microfarmers use waste streams to cultivate algae and harvest the biomass to make food, feed, fuel, fertilizer, and other valuable coproducts.

Microfarmers cultivate algae and possibly other microorganisms as they follow one or a combination of four sustainable and affordable food and energy, (SAFE) production paths, Figure 4.1.

Figure 4.1 Abundance SAFE Production Paths

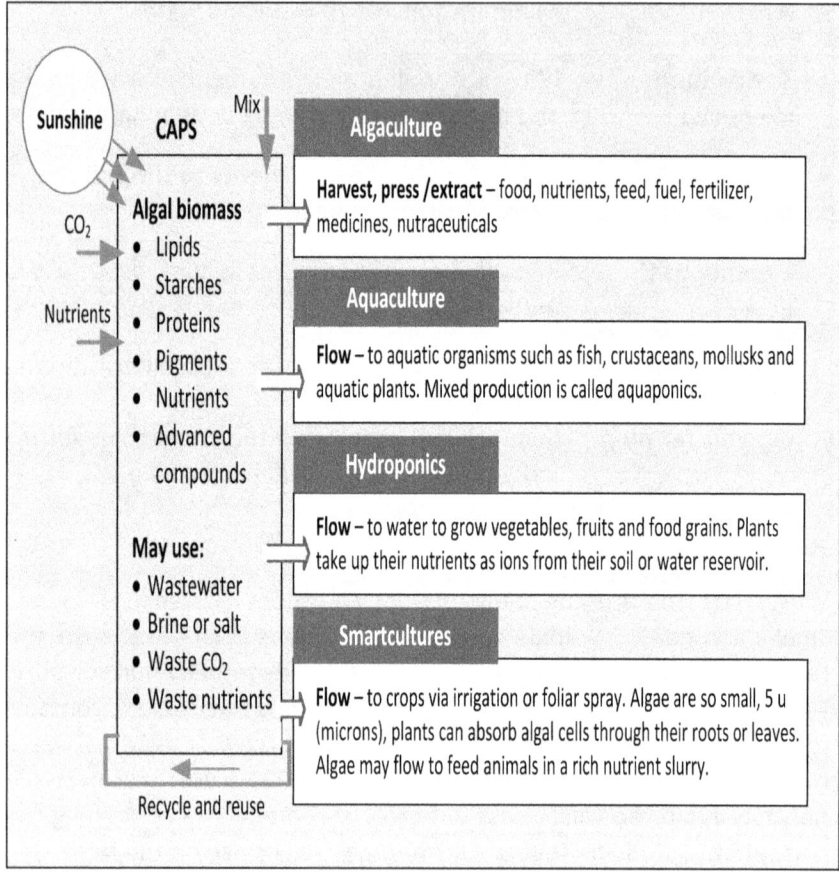

Algaculture

Algaculture grows microalgae or macroalgae, (seaweed), for commercial purposes or domestic needs. One-third of the algae grown commercially currently go to feed fish and shellfish. Extraction of the algae biomass enables the farmer to use the energy, nutrients and various coproducts for local needs, Table 4.1. Algaculture producers use many CAPS shapes, sizes and forms including ponds, troughs, semi-closed, and closed systems. Food, health food, feed, and nutraceutical producers include Earthrise, Phyco Bioscience, Algae BioSciences, Solazyme, Desert Sweet Biofuels, Herbules Pharma, Truly Natural, and Martek Biosciences.

CAPS may be sited near a carbon source such as a waste pile on a farm or a coal fired power, cement or manufacturing plant. Others may site cultures near a waste treatment facility. Some farmers may need to source organic wastes locally to supply sufficient nutrients. Algae grow well in fresh water but communities have competing needs for sweet water. Many communities have substantial sources of gray water that is not potable but excellent for growing algae. Some farms have reservoirs, ponds or wetlands that capture farm runoff, which are perfect for growing algae.

Half of the water stored in the earth's crust is brine water that is too salty for human use for irrigation but algae thrive in it. Many deserts, including those in the US Southwest have huge underlying oceans of brine water in relatively shallow aquifers which could produce millions of tons of algae biomass for food and biofuels.

Growing algae as fodder for animals, birds or aquatic creatures will be popular in many settings because animal fodder requires lower levels of cleanliness (except for pets) than producing food for direct human consumption. Waste water CAPS can produce food quality algae with the proper mechanisms in place. Considerable wastewater production may be used as feed or fertilizer. Many human grade valuable coproducts can be extracted from wastewater algae such as vitamins, minerals, antioxidants, trace elements pigments, oil and carrageen. Carrageens provide a wide assortment of uses in the food and other industries as thickening and stabilizing agents, as well as emulsifiers.

Table 4.1 Algaculture for Food, Biofuels and Novel Solutions

Food	Biofuels	Novel Solutions
Primary • Protein • Lipids – oils • Carbohydrates • Nucleic acids **Secondary** • Flour • Meat enhancer • Ice cream • Milk substitute • Sugar substitute • Sea vegetables • Food ingredients • Emulsifiers and thickeners • Novel flavors and textures • Pigments • Health foods • Nutraceuticals • Omega 3s **Feed and fodder** • Pets, fish, fowl • Meat animals • Micronutrients • Medicines and vaccines	**Primary** • Gasoline • Clean diesel • Methanol/ethanol • JP-8 jet fuel **Secondary** • Aviation gasoline • Alcohols • Hydrogen • Asphalt • Plastics, biodegradable • Rubber substitute **Biofertilizers** • Organic N-P-K • Bioavailable target nutrients • Micronutrients • Plant hormones • Soil organics • Build soil structure • Improve porosity • Plant growth regulators • Natural pesticides • Natural herbicides	**Air** • Carbon sequestration • Carbon capture/recycle • Capture sulfur • Capture heavy metals **Water – clean** • Waste streams – municipal, industrial, farm, brine, and ocean • Recover heavy metals **Cosmetics** • Moisturizers • Skin care **Local algae production** • Foreign aid • Disaster relief • Hunger and poverty **Medicines** • HIV / AIDS and SARS • Vaccines • Antibiotics /antiviral • Burns and bruises • Stomach remedies • Anti-cancer toxins • Pharmaceuticals • Advanced compounds

Hydroponics

A portion of the algaculture flows to containers where vegetables, grains and fruits grow in the rich algae water. Algae provide all of the macro and micronutrients necessary to grow large, colorful and tasty produce. Farmers have been using smelly fish for decades to improve plant germination, growth, and yields. Algae provide a less expensive alternative with a better nutrient profile for plants than fish oil with a pleasant organic smell.

Hydroponic farmers grow plants using mineral solutions in water rather than soil. In natural conditions, soil acts as a mineral nutrient reservoir but the soil itself is not essential for plant growth. Terrestrial plants may be grown with their roots in an inert medium such as perlite, gravel, mineral wool or nutrient solution. Hydroponic crop yields we found to be no better than crop yields with good soils because crop yields were ultimately limited by factors other than mineral nutrients, especially light. Later research showed that hydroponics offers other advantages, including constant access to oxygen, and that the plants have access to as much or as little water and nutrients as they need. Hydroponics producers have successfully grown vegetables on volcanic islands that lack fertile soil.

Growing algae to feed vegetables in water – hydroponics

Aeroponics, developed largely by NASA for space travel, grows plants in an air or fine mist environment without soil or aggregate medium. Aeroponics culture differs from both hydroponics and *in-vitro* production (plant tissue culture). Unlike hydroponics that uses water as growing medium and essential minerals to sustain plant growth, aeroponics cultures grow without an aggregate medium. Growth nutrients are transmitted by water mist, so aeroponics is often considered a form of hydroponics.

Aquaculture and Aquaponics

Aquaculture farmers grow fish and shellfish that feed on aquatic plants such as algae. Algae represent the preferred diet for most fish fry (immature fish) because the cells are small enough for the fry to eat. Most fish evolved on an algae diet in their natural settings. Most fish grow faster and have fewer digestive problems on algae compared with food grains.

The Chinese have practiced aquaculture since 2500 BC. Today, half the world's commercial fish and shellfish production comes from aquaculture. A recent scientific study reported that over 90% of the large fish have been extracted from the oceans. Unfortunately, many of the smaller fish have also been substantially over harvested. With diminishing natural fisheries in oceans, rivers, lakes, and estuaries, aquaculture will play a larger role in our food supply.

Growing algae to feed fish – Aquaculture

Aquaponics integrates fish and plant farming. Farmers grow algae to feed fish that add urea to the water. The nitrogen rich water flows to hydroponic greenhouses where vegetables and fruits grow in the high nutrient water. Polycultures can grow food with renewable energy and in closed systems, minimize consumption of fossil resources, including power, and fresh water.

Smartcultures

Smartcultures reengineer our food production system, beginning at its foundation – soil – with tiny microflora attracted by algae in plant roots that are ingeniously self regulating and self regenerative. Smartcultures move farmers toward abundance production by significantly reducing but not eliminating the use of fossil resources for growing field crops. Farmers using smartcultures are able to leave every field better than they found it.

Sustainable MicroAlgae Regenerative Technologies, (smartcultures), enable dirt farmers to recover, recycle, and reuse the energy and nutrients in their farm's waste stream to improve crop quality, taste, and yields while reducing operational costs by 30 to 50%. Smartculture farmers skip the harvest step and simply "grow and flow" the algae culture directly to their fields to recycle organic fertilizer to their crops that is immediately bioavailable to the plants. Smartcultures deliver 74 nutrients and trace elements that plants use to grow faster and stronger, and produce higher yields.

Smartcultures grow algae in the farm waste stream to fertilize fields

This portable CAPS can be moved to various waste sources. Design variations enable several configurations, depending on the setting. Smartculture reuse farm waste energy and nutrients.

Abundant Agriculture

Farmers who practice Smartcultures methods may:

- Increase income from 20 to 50% by improving crop yield and quality – micronutrients, vitamins, antioxidants, color, taste, texture, and shelf life.
- Save money by lowering fuel consumption by 20 to 30%.
- Save money by decreasing chemical fertilizers by 30 to 50%.
- Reduce air, soil, and water pollution by 80 to 95%.

Farms without irrigation systems spray the algae solution on the fields. The algae not only provide an organic fertilizer delivery system directly to the roots of crops but algae continues to grow in the field, as long as moisture is present, regenerating soils, and creating additional organic material. Algae's ability to extract in situ nutrients provides a tremendous advantage. Most farmers have available waste streams from human, animal, and vegetative wastes on which algae can thrive. Rather than spending 30-40% of their production costs on fertilizer, algae may cut nutrient cost in half because algae can recover 90% of the nutrients from the farm waste stream.

Smartcultures employ a set of technologies that imitate nature to provide enhanced foundation (soil structure) and food (nutrients) to plants. Every farmer and gardener knows plants thrive in amended soils; they grow faster, stronger, and larger, and they have better taste, and texture.

Smartcultures transform agricultural methods. Rather than mining and paying high prices for chemical fertilizers and using them once, farmers can continuously recycle and reuse nutrients from the farm's waste stream. Smartcultures reverses systemic extraction of soil nutrients and organics because growers can cultivate algae and microflora to add nutrients and organics in the field. Smartcultures avoids the use of chemicals that destroy soil microbes and soil structure as the microbial communities act synergistically to improve soil structure. Industrial agriculture degrades soil, promotes erosion and creates severe pollution. Smartcultures improve soil structure and reduce nutrient waste, erosion, and pollution.

Recent fertilizer price escalation has pushed the cost of fertilizer to 30 to 40% of farm operating costs. The cost of P fertilizer, for example, rose 700% in a recent 14 month period. Farmers face two serious problems with industrial fertilizers: bioavailability and erosion. Chemical fertilizers must first be broken down by microorganisms, e.g. algae in the soil before they can be absorbed by the plant. The process may take months or years so farmers have to put on far more fertilizers than the plant needs in order to maximize growth. Much of the applied fertilizer does not reach the crop and erodes with irrigation, winds, and rains. The next year, the farmer must apply even more fertilizer to achieve the same yields. This model is sustainable only as long as fertilizers are cheap and soils do not wear out.

Smartcultures enable farmers to minimize the high cost of fertilizers by recovering nutrients from the farm waste stream. Farmers can improve the quality and quantity of field crops because algae biofertilizers are immediately bioavailable to the plants and create almost no waste. Target nutrients carried in algae biofertilizers can be delivered in precise amounts at specific times during a crop's growing cycle – which can maximize germination, early growth, maturation, and fruiting.

Algae cultures naturally bioaccumulate the full spectrum of macro and micronutrients crops need. CAPS near fields can overload one or more nutrients and deliver them to crops exactly when needed. For example, adding more calcium when the crop is fruiting may enhance fruit size, weight, color, texture, and taste.

Farmers can save money and energy by lowering their consumption of fuel due to easier cultivation. In some settings, smartcultures have improved soil porosity (loosen compacted soil) by 500%. Recovering and recycling organic fertilizers from the farm waste stream can reduce the use of chemical fertilizers substantially. Smartcultures reduce air, soil, and water and pollution because algae biofertilizers and plant growth hormones significantly diminish the need for agricultural chemicals.

Drip irrigation can deliver algae biofertilizers precisely to the roots, minimizing water use and the waste of nutrients. Algae continue to grow in the soil while moisture is present, which adds rich organic matter and conditions the soil, making it more erosion resistant. This model may also use no or minimal till to minimize soil disruption and provide longevity to the water-efficient drip irrigation system. In non-irrigated settings, the algae culture may be applied with a field or aerial sprayer.

Several other agriculture methods move farmers toward sustainable production.

Organic Farming

Peter Fossel, Raoul Adamchak, the Rodale Institute and a chorus of others recommend a return to organic farming. Organic farming has been practiced successfully for centuries. In contrast to the Green Revolution's reliance on synthetic N, inorganic P, and K to maximize production, organic farmers use crop rotation, green manure (cover crops), compost, biological pest control, and minimal cultivation. These actions help to maintain soil productivity and control pests. Organic production limits the use of synthetic fertilizers, agricultural chemicals, livestock feed additives, and GM seeds.

If organic farms could produce enough food to supply current populations, farmers would quickly adopt organic methods. In spite of the numerous calls for organic food production, less than 1% of the world's croplands are farmed organically. In 2007, 4% of the European Union's farms (where farmers receive subsidies to use organic methods) were organically managed compared to 0.6% of US farmland.

Organic farming is labor and knowledge-intensive whereas conventional farming is capital-intensive, requiring more energy, manufactured, and external inputs. Organic farmers use labor and ecological knowledge rather than chemicals to manage and control weeds and pests. Since organic farmers cannot leverage external and synthetic inputs as efficiently, they are less able to achieve economies of scale and are typically far smaller farms than industrial farms.

Organic farming moderates but does not solve the fossil resource consumption problem. Organic farmers typically consume more cropland, fresh water and fuel per calorie because they must produce both compost and food crops; and yields may be lower. One of the main choke points for organic farmers is that organic fertilizers may be more expensive than inorganic. In many places, organic fertilizers are simply unavailable. Grain farmers in the US commonly find no organic fertilizer available locally because cattle feeding operations are in Texas, thousands of miles away from food grain fields in Iowa.

Sustainable Agriculture

The National Sustainable Agriculture Coalition describes sustainable agriculture as: "A safe, nutritious, ample, and affordable food supply that is produced by a legion of family farmers who make a decent living pursuing their trade, while protecting and improving the environment and contributing to the strength and stability of their communities." The Brundland Report states that sustainability "meets the needs and aspirations of the present without compromising the ability of future generations to meet their own needs."

Sustainable agriculture integrates three main goals: environmental stewardship, farm profitability and prosperous farming communities. Sustainable agriculture came about in the 1960s as farmers and environmentalists became concerned about the impacts of agricultural chemicals on the natural environment and human health. Thomas Lyson adds a local perspective that nurtures local community development that he calls civic agriculture. Leo Horrigan and his team at Johns Hopkins, at the Center for a Livable Future extended the definition to include more equitable distribution of high-protein food; without animal fat.

John Ikerd in *Crisis and Opportunity: Sustainability in American Agriculture* eloquently describes stainable agriculture as a movement by small farmers that may claim a label of organic, low-input, holistic, practical or real farmer. Sustainable farmers try to build farming systems that are ecologically sound, economically viable, and socially responsible. Sparse empirical research exists on sustainable farming independent of the research on the benefits of organic production.

Conservation Agriculture

Several international R&D organizations advocate conservation agriculture (CA) and claim the practices are a panacea for regions with poor agricultural productivity and soil degradation such as sub-Saharan Africa. CA replaces tillage with a heavy dependence on fertilizers, herbicides and pesticides. Proponents claim CA increases yields, reduces labor, improves soil fertility, and reduces erosion. Unfortunately, empirical evidence shows mixed results, and it is not clear what principles of CA contribute to which effects.

Permaculture

Permaculture is a whole systems design to permanent agriculture. Practitioners design human settlements and agricultural systems that align with the relationships found in natural ecosystems with the intent to be self-sufficient. Self-sufficiency reduces the reliance on industrial production and distribution systems that are systematically destroying Earth's ecosystems. Unlike industrial agriculture, permaculture farmers use minimal external inputs and grow diverse crops. Small-scale food production, local markets, and minimal food miles provides a permaculture model.

Vertical farms

Dickson Despommier's book, *Vertical Farms* describes climate controlled greenhouses stacked vertically. While no vertical farms currently exist, potentially they could produce climate independent crops using minimal fossil resources, while creating no external pollution. Vertical farms built in cities would cut food transportation cost substantially.

Vertical farms could grow considerable produce but the construction and operating costs have not been calculated but obviously substantial. Vertical farms would tend to concentrate food production in the hands of those with sufficient wealth to construct vertical farms, which would have an adverse impact on social equity. Vertical farms would provide an excellent growing model for abundant agriculture if the cost structure can be managed.

Chapter 5. Abundance Microfarms

He that will not apply new remedies must expect new evils; for time is the greatest innovator.

– Sir Francis Bacon

Growing algae for commercial or local consumption purposes, algaculture is a nascent industry that is primed to grow exponentially. About 150 commercial algae producers cultivate algae in 2010, primarily for advanced compounds such as Omega 3 fatty acids, nutraceuticals, and health foods.

The key to algae production economics lies in sourcing affordable inputs, which is why 360 microfarms will be so effective. Farmers who cultivate algae commercially today use primarily fossil resources. Producers who maximize the use of renewable, surplus or waste inputs that are sustainable, practice abundance. Fortunately, most algae producers are frugal and look for surplus or free inputs wherever possible. Some producers also employ renewable energy for algae cultivation, harvest, and extraction.

_____

Disclosure: The author serves on numerous non-fiduciary boards for algae companies, including most the firms profiled here to illustrate various abundance production models.

Natural stands

Algae grow quickly in natural settings such as wetlands, creeks, lakes, ponds, pools, and puddles. Gardeners can harvest algae from natural stands using a bucket. Irrigation canals often have an algae crust left after the water recedes which is easy to scrape into a container. Algae concentrates nutrients from the irrigation water and makes those nutrients immediately bioavailable to plants.

Gardeners can create a poor man's smartculture system by placing an old aquarium or any other plastic container filled with water in the sun. Throw in some garden or kitchen wastes for carbon and nutrients and algae will grow quickly and provide excellent fertilizer. A bubbler mixes the water and speeds the culture growth but is not required. Homemade systems require cleaning the inside regularly to allow sufficient light to reach the the culture.

Algae often grow in mats along the top of the water column, giving the cells access to light. These mats are easy to harvest by skimming the water. Streams often have filamentous algae that grow in the fashion of ropes. Algae strings can be pulled from streams easily and provide rich organic material to the soil, along with the full set of nutrients, minerals, and trace elements that enable plants to flourish.

Cultivation in ponds

Algae pond systems in the US were first developed for water treatment. The recovered biomass was converted to methane and burned as a local source of energy. When fuel was cheap, the energy value from algae was considered incidental.

Most microalgae production today occurs in open ponds, called "slime ranching," because ponds are cheap to build and operate. However, open pond cultures make it difficult to control invasion from weed algae and microorganism predators such as ameba, ciliates, bacteria, rotifers, viruses, fungi, and zooplankton. Open ponds are vulnerable to contamination from dust, windborne organisms, insects, and birds. In addition, ponds are limited geographically to tropical and subtropical areas with warm temperatures, low rainfall, and little cloud cover.

Algae Grown in Ponds

Evaporation represents a critical limitation to open ponds, especially in hot climates. An open pond may lose half its water in a growing season which equates to water loss similar to irrigated crops. Evaporation increases retained salts, which may affect the stability of the culture. Some open ponds use seawater, waste or brine water which often makes the water free but does not slow evaporation or salt concentration.

Most microalgae need light and carbon dioxide but they vary substantially by specie in nutrient and environmental requirements. Some species flourish in unlined ponds in Australia but the same variety may grow poorly in unlined ponds in India or China. Consequently, local conditions often dictate the design and construction of open ponds as well as species selection, and cultivation methods.

Algae ponds typically are shallow, one to three feet deep, in order to maximize cell access to light. Algae grow quickly and new cells shade older cells so in unmixed ponds will have growth only in the top two inches of the water column. The ponds are typically mixed with a paddle wheel or other mechanism that keeps the culture moving around the raceway in a circle. Water movement around the raceway creates enough turbulence to bounce cells to the surface where they can absorb photons. Large ponds used for municipal water remediation bubble a mixture of CO_2, and air to move the water.

Large ponds tend to be less expensive per unit area to construct and operate. Size affects water circulation, operating costs, mixing systems, and species selection. Mixing gives cells access to light, prevents cells from settling to the bottom, and avoids thermal and oxygen stratification in the pond. Effective mixing systems increase cell density, which reduces harvesting cost. Harvesting typically occurs with centrifuge, filters, or flocculation.

Algae Cultures in open Raceways

Commercial algae production in outdoor ponds is vulnerable to weed algae invasion from competing local species. Producers grow *Spirulina* at high bicarbonate concentrations, which yield a high pH that discourages competing species. *D. salina* is grown in a high saline culture that minimizes contamination by unwanted algae species. Outdoor ponds often use a batch culture approach to control contamination where the culture is restarted at regular intervals with a fresh single-specie culture.

Pond and trough production models

The variety of successful algae production models using simple ponds and troughs demonstrates the potential for innovation in algaculture. The first model overcomes the cost hurdle with a $10 solution.

Patrick Hallman, program chair for the Rare Fruit Growers, fills two play pools with water, which attracts ducks to his backyard. The ducks do their business in the pool, which provides nutrients and carbon for algae growth. Every week or so, Patrick takes a bucket and uses the algae water to fertilize the family garden. He says the water becomes a compost tea that works as great fertilizers for pomegranates, sugar cane, citrus and other fruit trees.

Smartcultures speed up the process of composting. Composting piles take several months or years to break down. Manure or shredded botanical waste can be broken down by algae in a few days. Once the organic matter has been absorbed by the alga cells, the nutrients are carried by algae, which are 5 μ small. These small cells are immediately bioavailable to plants when they are delivered to the roots or foliar, on leaves and stems.

Patrick Hallman's simple Duck Pond creates Compost Tea

Human mothers prepare food in a similar fashion for their infants. Mothers mash the food so it fits in the child's tiny mouth. Algae play a similar maternal role for plants.

Earthrise Nutritionals is the longest running commercial algae producer in the US. Earthrise CEO Amha Belay edited an excellent book on spirulina production and marketing, *Spirulina in Human Nutrition and Health, (2006)*. A prior Earthrise CEO, Robert Henrikson also wrote an exceptional book on microalgae and health, *Spirulina: World Food*, (2010.) Earthrise provides a set of tasty algae food recipes at http://www.earthrise.com/reviewRecipe/recipe.html.

Earthrise Nutritionals located in the Sonoran Desert of Southeastern California produces 500 tons of food-grade Spirulina in their 30, one acre, 18 inch deep Spirulina ponds. Their seven month production year grows Spirulina for health foods and nutraceutical markets. They could produce a cold season algae variety but do not currently.

Earthrise uses fresh water and industrial CO_2, and nutrients, which are added continually to support the algae. A 50-foot paddlewheel mixes and moves the algae culture around the raceway. This highly productive facility uses 30 year old technology. Earthrise has provided the industry with many insights and innovations, especially in producing food-grade spirulina.

Algaculture for Food at Earthrise Nutritionals

The **Laboratory for Algae Research and Biotechnology, LARB**, at Arizona State University Polytechnic campus, directed by Professors Milton Sommerfeld and Qiang Hu, operate a variety of algae growing systems in order to develop production metrics using open and closed systems. While the lab primary focuses on jet fuel production, scientists are examining every element of algae production in order to optimize sustained culture stability and productivity. LARB offers services that enable entrepreneurs to test their algae production design ideas against open and closed CAPS with sustained production metrics.

This simple and inexpensive open raceway with two baffles, cycles and mixes the algae culture using a paddlewheel. The baffles keep the water flowing evenly around the raceway. This is a fourth-generation CAPS may be scaled up or down. Half this culture is harvested daily. Open systems like raceways are easy to build and inexpensive to operate and maintain. They are good for growing a wide variety of algae species but are prone to contaminates, which require the culture to be changed and re-inoculated every month or two.

LARB Open Raceway

Desert Sweet Biofuels has been growing algae for 24 years to feed shrimp grown in one acre ponds. Gary Wood, CEO notes that every shrimp farmer is first an algae farmer because they grow algae to feed the shrimp. Gary Wood is in the process of transitioning the famous Desert Sweet Shrimp Farm in Gila Bend, Arizona to also grow algae for food, cattle feed, and biofuels.

The shrimp ponds are much deeper than typical algae ponds but innovative mixing systems keep the ponds productive. Temperatures in Gila Bend can reach 120° F in the summer and deeper ponds mitigate heat. All the ponds are plumbed with connective piping and the culture can be transferred, emptied or refilled in a single day.

Full and Empty Algae / Shrimp Ponds

Production begins with an inoculate culture in the runways below inside a greenhouse that is covered in the winter. Once inoculated has grown to sufficient density, part of the culture flows to other tanks that hold the shrimp fry. Shrimp grow rapidly on an algae diet and they are moved to ponds where they mature.

The other part of the inoculant culture flows to ponds where algae biomass grows. In a few weeks, the pond algae have sufficient density that half the biomass can be harvested daily. The harvester unit below filters the algae onto a bed where it is removed and dried in the sun.

Carbon Neutral Food Production – Desert Sweet Biofuels

Gary Wood grows algae for carbon-neutral food, feed, nutrients, and energy production. He grows algae with carbon and nutrients from the waste stream of the Wood Brothers farm. The carbon source comes from a pyrolysis unit on the far right that burns farm wastes in a closed kiln. The resulting pure H_2 drives an electric generator on the left that in turn powers the pond mixing system.

The CO vents to the algae pond through the tubes stringing into the water where it feeds the algae. The biochar residual remaining the pyrolysis unit provides a slow release fertilizer on the farm's alfalfa fields, and olive groves. Most farms generate sufficient botanical and animal wastes to use the abundance model for algae production.

Runway within a Greenhouse and Algae Filter Harvester

The production of algae biomass is carbon neutral because no fossil carbon is released to the atmosphere. Of course, the carbon from farm wastes is released (recycled) with the consumption of the food, feed or fuel, but not the fertilizer. Analysis shows that the biochar sequesters 10 to 20% of the total carbon dioxide.

Gary Wood visualizes cooperatives joining 10 to 20 farms in each area, much like cotton gins. Farmers will cultivate and harvest algae on their farms. A co-op truck will come by several days a week to pick up the algae and carry it to a central processing facility for component extraction. The resulting algae-based products might be used by the co-op members or sold on the open market.

Desert Sweet Biofuels Algae Pond and Mixing Propeller

Algae serve as a remediation tool for wastewater and can remove N, P, and K in livestock manure. Walter Mulbry, an Agricultural Research Service scientist set up four algae turf scrubbers outside dairy barns.

The shallow 100-foot raceways were covered with nylon netting that created a scaffold where the algae could grow. For the next three years, from April until December, a submerged water pump at one end of the raceways circulated a mix of fresh water and raw or anaerobically digested dairy manure effluent over the algae. The raceways supported thriving colonies of green filamentous algae.

Algae growing in a Turf Scrubber and Air Dried

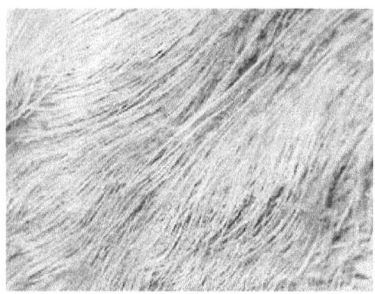

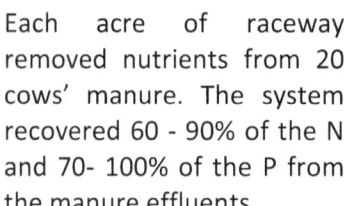

Each acre of raceway removed nutrients from 20 cows' manure. The system recovered 60 - 90% of the N and 70- 100% of the P from the manure effluents.

Mubry's research team calculated the recovery cost was comparable to other manure management practices—with a per pound cost of $5 for N, and $25 for P. Dried algae made excellent organic fertilizer as corn and cucumber seedlings grown in algae-amended potting mixes performed as well as those grown with commercial fertilizers.

Ben Cloud, CEO leads **Phyco Biosciences** based in Phoenix, Arizona where they use lined troughs to produce algae. Phyco's Super Trough algae production system represents a low cost, large-scale technology for algae cultivation. Ben Cloud says the Super Trough provides a 30% reduction in capital cost and 50% reduction in operating cost when compared to open ponds and raceway systems.

The cost advantages of the Super Trough are derived from the ability to install 1,250 feet of Phyco's liner system in a single pass using a tractor, dramatically reducing installation time and cost while enabling rapid scalability. The trough liner material contains an integrated liquid and gas emitter system capable of distributing crop inputs uniformly over 40 acres in a matter of minutes. The ability to manage nutrients enabled by the integrated distribution system provides significantly higher yields than competing production technologies at a lower cost. The system enables better culture control and operator versatility which improves algae production.

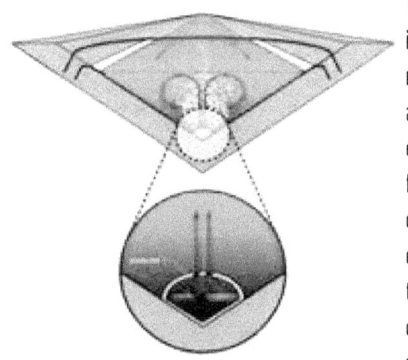

The Super Trough liner system includes an aeration emitter made from a DuPont material, and a second special nutrient emitter. Economical fabrication, fast, mechanical installation, and optimal operating conditions created by the "V" shaped trough and uniform distribution of CO_2 gases and nutrients assure rapid culture growth and stability.

Super Trough Air Flow

From an operator perspective, large-scale production is achievable due to effective control of the algae culture over large areas.

Phyco Biosciences Super Trough – Simple Installation

The key Super Trough advancements include low cost but durable components, and uniform distribution of liquid and gaseous inputs. The Super Trough system uses little energy for the air drive system that circulates the culture on a level 40 acre field.

The trough surface area is 60" wide and 18" deep. The algae fields are leveled with common agricultural leveling equipment. A tractor creates the furrows and then lays the liner. Dirt holds the liner in place. A 40 acre system can be laid out in a few days. The liners are designed to last 10 years. The troughs can be mechanically covered with a disposable solar cover to retain heat in colder climates or to extend the production season

Phyco Biosciences Super Trough Growing Algae

The trough design uses 60% less energy compared with paddle wheel systems. The aeration emitter continually rolls the culture giving all the cells regular access to light. Aeration bubbles are visible in the picture above. Preliminary field production indicates the Super Trough system out produces ponds or raceways, with a lower operating cost. The system design enables farmers to grow a wide variety of algae species. Some business plans grow three species a year, one each for summer, winter and spring/fall.

Phyco Biosciences has also developed a low-cost, commercial scale harvester designed specifically for dewatering and drying algae biomass. Focusing on simplicity, automation, and high volume capacities, this technology efficiently separates a wide variety of algae species from large volumes of water using a fraction of the energy of conventional dewatering and drying systems. This technology is a result of a collaborative effort between Phyco Biosciences and Algaeventure Systems.

Phyco Biosciences Harvester and Ron Henson with Spirulina

Algae Harvesters – Drum and Flat Bed

Closed CAPS models

Semi-closed or closed CAPS are containers designed to capture maximum sunshine and keep the culture stable and secure. CAPS vary in size from a several square meters to many hectors. Growing containers provide considerable visual variety and may be covered ponds, plastic bags, plastic sheets, resins, or glass – anything that allows light to penetrate. Some systems use fiber optics or mirrors to capture additional sunlight or add artificial light.

Algae growing systems are commonly called photobioreactors which our consumer research indicates creates negative consumer perceptions and feelings. Even though the term photobioreactors implies that the sun excites plant cells to produce biomass through photosynthesis, naïve observer's associate reactors with nuclear power. Additionally, the term bioreactor has become synonymous with garbage waste disposal. Consequently, the terms used here are solar garden, biofactory or cultivated algaculture production system.

Closed CAPS producers select algae species that maximize the characteristics desired such as biomass percentage of lipids, protein, or component product. Food growers select species to maximize biomass protein while biofuel producers select a species with high lipid content.

Algaelab.org

Aaron Wolf Baum, aka, Dr. Friendly at algaelab.org offers a hands-on home-grown algae workshop series. For a $350 fee, participants can take home a personal-sized and growing system that can be placed in a sunny window. The CAPS can provide enough spirulina to supplement the diet of one person every day. Larger installations can provide food, feed, organic fertilizer and other forms of energy.

The workshop includes a discussion of the practicalities of setting up a microfarm. There is an extended hands-on section of the workshop where Aaron shows participants how to set up and run the personal photo-bioreactor kit. He promises that it's easy. Aaron also provides a liter of live spirulina for a starter culture. Full information and workshop schedules are available at: www.algaelab.org.

Do-it-Yourself CAPS Kit from Aaron Wolf Baum

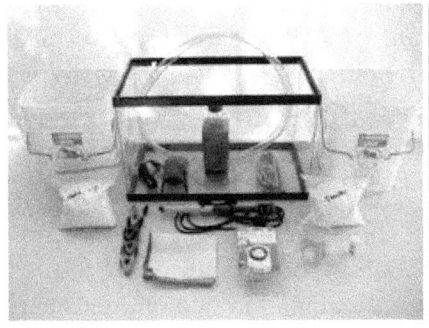

The term "closed system" is a misnomer because algae predators and weed species invade any growing system. It may be better to think of a closed system, as an arrangement that gives the producer more control over production parameters. Compared with open CAPS, growers have more but not perfect control over contamination from invasive algae and predators. Farmers typically control predators and weed species with a combination of parameters including pH, temperature and nutrients.

Acrylic that is ultraviolet (UV)-stabilized is typically used for CAPS construction because it is cheaper, stronger, lighter, more flexible, and easier to fabricate than glass. Assembly of the biofactory requires the integration of the various mixing, monitoring, and controlling subsystems.

Tubular CAPS and Plastic Bags

The **Laboratory for Algae Research and Biotechnology, LARB** at the ASU Polytechnic campus produces algae for jet fuel in rows of plate CAPS. The plate containers are 4 feet by 4 feet and only 4 inches think. Tests with hundreds of different growing systems demonstrated that the highest production occurs with relatively thin containers. Growing systems have a vertical limited because algae is soft and too much water pressure causes shear stress.

Air and CO_2 are bubbled through the container to constantly mix the culture. These CAPS also have a mist system to cool the culture in the Arizona summer. The mist water that runs down the side of the CAPS, similar to the culture water, is continuously recycled through an underground cistern.

Rectangular Plate CAPS – Arizona State University

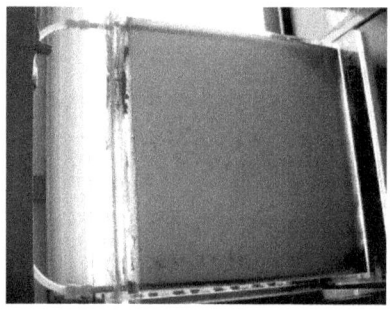

These plate CAPS not only optimize production but minimize operational costs, including cleaning and maintenance. The field site is operational 330 days a year with just a few weeks of downtime for CAPS maintenance. These CAPS are modular and designed to be scaled larger or smaller to meet production needs.

Closed systems offer far more control over growing parameters than open ponds or natural settings. Stressing algae to create more production of desired components by changing production parameters is practical only in closed systems. Closed systems also avoid water loss from evaporation.

A2BE Carbon Capture of Boulder, Colorado builds carbon capture and recycle, CCR, systems that take advantage of algae's capacity to profitably recycle industrial CO_2 emissions into fuel and other coproducts. Mark Allen, CEO, says their advanced energy-conversion system combines algal CO_2 capture technologies with biomass gasification and creates an integrated renewable fuel production system. The CO_2 can be recycled from any source and the biomass feedstock for gasification into syngas may come from wood waste, municipal solid waste or the processed algae waste. The CO_2 produced from the biomass gasification process is recycled to grow more algae.

Jim Sears, President and CTO developed the patented system design and notes that the CCR biofactories are scalable from a few acres to large farms that recycle industrial CO_2 emissions into algal biomass that can provide a wide range of valuable products.

At the core of the technology is the algae growing and harvesting CAPS. Each machine is 450' long and 50' wide consisting of twin 20' wide x 10" deep x 300' long, transparent plastic algae water-beds. It holds 150,000 gallons of algae. The biofactories work with any species of algae including cyanobacteria and diatoms.

A2BE's Carbon Capture and Recycle Biofactory

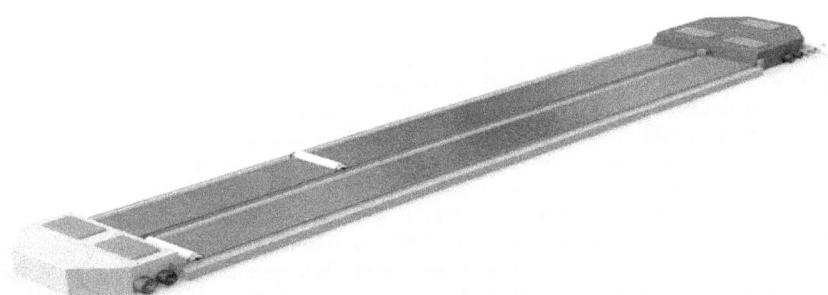

The harvesting technology is adaptable to fit local needs. A2BE offers a novel bioharvesting technology where brine shrimp feed on the algae and the shrimp are harvested and processed. The CCR machine is climate adaptive due to thermal barriers above or below the culture flow that regulate temperature. This allows deployment nearly anywhere with sunshine.

The A2BE business model shows how CO_2 recycling is profitable. Their business plan shows each ton of CO_2 may be captured at a cost of about $40 for nutrients and $10 for the CCR biofactory and operations. The net revenue of $200 per ton of CO_2 captured is based on: oil ($40), protein ($90), methane ($25), fertilizer ($40), oxygen ($30) and CO_2 credit ($25).

A2BE has created an even more compelling production attribute than the profitability per ton of CO_2. Their CCR biofactory creates a carbon negative process because each ton of carbon captured and recycled into the various algal coproducts displaces and avoids about 1.25 tons of carbon entering the atmosphere. The carbon negative process holds true when the original carbon is fossil sourced and the resulting products are burned as fuel.

A2BE's Vision of a large Algae Farm

A2BE is not only building a company to take on the substantial challenge of carbon capture but they are building a collaborative group of select institutions, corporations, and key researchers to address the spectrum of talents and disciplines needed to rapidly commercialize a solution called algae@work.

Greenwater Global produces algae anywhere including in buildings or barns using existing engineering methods, established technology, and readily available equipment components. CEO John Jackson says this indoor factory model offers the lowest cost and highest output of algae and its coproducts which enables farmers to achieve maximum profitability with no waste or negative environmental impacts.

GreenWater Global wants to make it possible to grow algae anytime, anywhere, 24 hours a day, 7 days a week, 365 days a year. The three 19,000 gallon tanks in the picture have yet to have the plumbing installed. Panels of grow lights set vertically in the tanks provide the light energy. Each tank has a full set of automated monitoring and diagnostic equipment, which can be viewed remotely.

John Jackson, CEO of Greenwater Global with Indoor CAPS

The Greenwater Global tanks are modular and can be assembled in a single day. Plumbing and attaching the monitoring equipment requires additional time. The tanks can be scaled to any size and placed indoors or outdoors.

Algae Biosciences produces algae using pristine brine water from an aquifer near Holbrook in northern Arizona. The fossil aquifer had no exchange with annual rains or runoff so it remains clean of pollutants.

Andrew Ayers, CDO says Algae Bioscience's consistent sunlight exposure, ultra-pure water supply and advanced technology enable production of twice as much algae as its competitors; at a lower cost.

The AlgaeBio business model focuses on supplying valuable ingredients to nutraceutical, pharmaceutical and health food companies. AlgaeBio produces high-quality products from various species of marine micro and macroalgae. AlgaeBio will produce nutritional supplement ingredients for human, animal, and aquaculture consumption. They will also produce cosmetics, fluorescent dyes, and a variety of other products in diverse markets.

Marine micro and macroalgae offer a wide variety of advanced compounds because there are thousands probably millions of species throughout the oceans. AlgaeBio can also adapt freshwater algae to brine water conditions, if the need is identified for particular compound known to be in freshwater algae.

Algae BioScience CAPS

AlgaeBio produces algae with a hybrid combination of closed and open CAPS. The closed system, shown on the right, inside the greenhouse, grows the inocula rapidly. Once the inocula have concentrated, the culture flows to outdoor raceways for mass production. The hybrid CAPS gives AlgaeBio high control over the sensitive inocula growth phase. The raceways can be covered in winter to retain heat.

AlgaeBio's produces at 5,000 feet elevation where they use a greenhouse to keep the initial cultures warm in the winter. The area gets 360 days of sunshine the year, which is great for algae production. The land sits over a huge brine aquifer and cannot support agriculture due to high soil salinity. Since nothing can grow on the land, land prices are very cheap.

Algae BioScience Greenhouse and Extraction Equipment

The greenhouse is designed to hold 3000 CAPS units. AlgaeBio harvests 50% of the biomass from the raceways daily with their proprietary extraction equipment.

Smartcultures

Nearly any CAPS model serves to cultivate algae for the purpose of fertilizing and regenerating fields or gardens. Abundance farmers who use smartcultures simply "grow and flow" the culture to their fields. The farmer's goal in algaculture is to maximize total biomass or the target compounds. Smartcultures do not have to be optimized for productivity, so less sophisticated CAPS may serve the fields. This portable unit on a trailer supported a field of about 200 acres.

Smartcultures produces rich nutrients in algae for special delivery to field crops. Smartcultures mimics and enhances nature as the farmer:

1. Cultivates a few alga cells from the field. Indigenous algae are the most robust because they have adapted to local conditions.

2. Grows a starter culture of robust algae cells, the inoculant, to sufficient density. When they can flow into the CAPS.

3. Grows and concentrate the algae in the CAPS. Algae absorb the nutrients.

4. Flows the nutrient-packed algae back to the field in the irrigation.

Algae Bioscience Smartcultures – growing algae for field crops

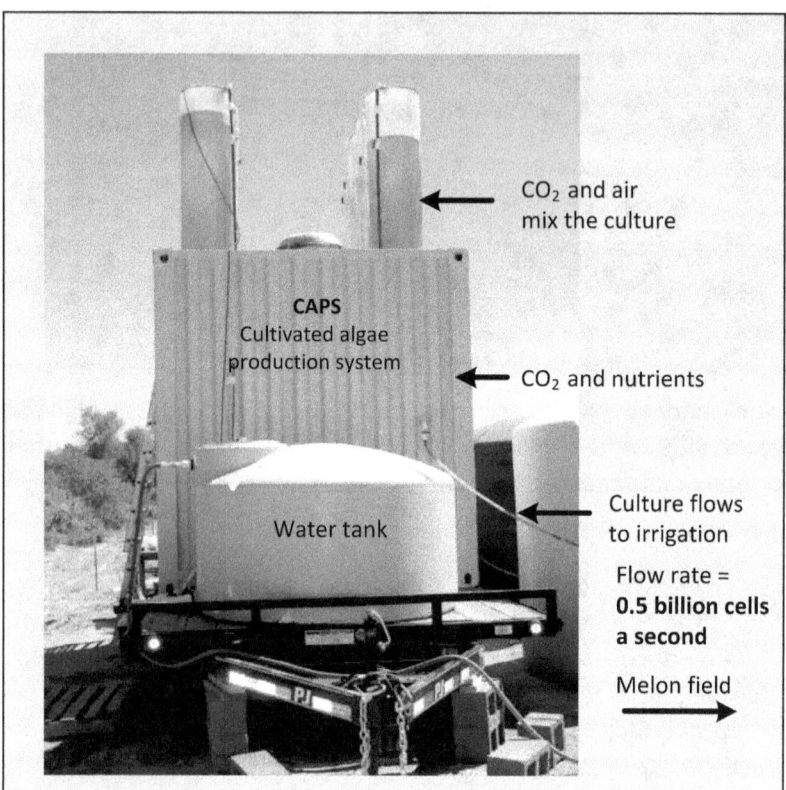

Algae carry organic energy, as well as 74 macro and micronutrients, vitamins and vital trace elements to crops. Plants respond quickly to the algae delivery system because the nutrients are immediately bioavailable. Crops grow faster and with more vitality that with chemical fertilizers.

Chemical fertilizers must be broken down by soil microbes, which include algae and the microorganism community. Plants are stronger and better able to withstand weather, wind, and pests. Algae transforms depleted soils to a rich, regenerated organic foundation that produces larger crops with better color, taste, texture, and restored nutrient profile.

Algae reduce irrigation needed by 30% by enhancing soil porosity, organics, and root depth. Fields supported with algae consume less fuel for cultivation because they are less compacted. Algae fed fields need less chemical fertilizer. Strong plants benefit from microflora that supplies natural growth hormones, which decrease the need for agricultural chemicals and poisons. Bioavailable nutrients in the algae are absorbed by the crops, which reduces fertilizer waste and erosion losses by 80%.

(I had not been called a swabby since Plebe year at Annapolis.)

Author swabbing a CAPS

Algae delivered in the irrigation water continue to grow in the soil as long as moisture is present. The algae not absorbed immediately by the plant add to soil organics, improve fertility and provide slow release nutrients similar to compost.

Algae-fed Melon and Control – Ripe, tasty Cantaloupe

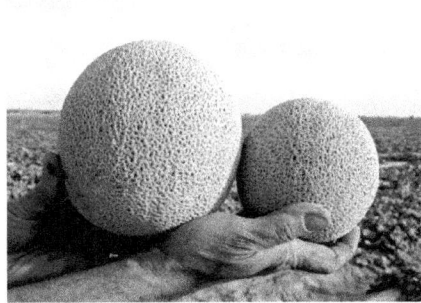 

Algae attract diverse communities of microflora including fungi, yeast, bacteria, viruses, slimes, and others that work symbiotically to produce compounds needed by plants. Algae grow polysaccharide sheaths that loosen the soil and enable the return of earthworms.

Scott Tollefson manages several farms for a multinational food company. His observation: "These are the largest plants and biggest fall melon crop I've ever seen. What's even more incredible, we've applied:

- 50% less N fertilizer (algae provide bioavailable N.)
- 40% less P fertilizer (algae provide bioavailable P.)
- 50% less diesel fuel (algae loosens compacted soil.)
- 60% less pesticides (algae stimulate natural plant hormones.)
- 50% less herbicides (algae strengthen plants to compete with weeds.)
- 66% less fungicides (specific algae attack fungi and nematodes.)"

"We've had severe white fly problems on our other fields. We have no whiteflies here and we haven't even sprayed. And this is only the first year with algae building soil fertility and capacity. Next year, I'm sure we'll get higher germination rates and a better crop. It all comes down to the soil. Algae enhance the soil and crops with a full set of nutrients and rich organic material."

The Algae Bioscience smartcultures process was repeated for a second year of spring and fall melon crops. The farm realized about a 30% improvement in yields with higher quality fruit. The algae fertilized field was ready for harvest nine days earlier than the control fields. The algae-infused fruit were larger, more colorful, and had about four days longer shelf life. A series of blind taste tests at Arizona State University showed a 16:1 preference for the algae-infused melons versus controls.

While the farm realized higher melon yields, the crop inputs decreased substantially. The crop needed 50% less N fertilizer because the blue-green algae delivered some N and fixes N from the atmosphere. Less P fertilizer was applied because green algae delivered some P and solubilized P locked in the soil. Algae increased soil porosity 500% because algae continues to grow in the soil adding soil organics, humus.

The Algal Biomass Organization, *Algae Industry Magazine*, and other networks serve nascent algae industry and the following firms.

Algae industry firms

Table 5.1 Firms working on Algae

Algae Alliance	Global Green Solutions
Algae BioFuels	GreenShift
Algae Biosciences	Green Start Products
Algadyne	GS Cleantech
Algaen	Heliae Development
Algenol	Infinifuel
Arare	Inventure
Aquaflow	Kent Sea Tech
Aurora Algae	Kiwikpower
BioProcess Algae	Live Fuels
Biodiesel	NanoVoltaix
Biofuels Digest	Open Algae
Biofuel Review	OriginOil
BioMat SA	Ocean Harvest Technology
Bionavitas	PetroAlgae
Blue Marble Energy	Plaatts
Carbon Capture Corp	Pelletbase
Cell Tech	Phycobiosciences
Diversified Energy	Raytheon
EnAgri	Renewable Energy Magazine
Energy Farms	Texas Clean Fuels
Energy Update	Sapphire Energy
Ethanol India	Simris Alg AB
Evados	Simplexity
Earthrise Nutritionals	Solazyme
Genergetics	Solix Biosystems
GreenEnergy	World Oil

Growing systems

Cultivated algae production systems, CAPS share common elements as shown in Table 5.2. The optimal shape, size, and configuration gardens are currently under investigation. In order to meet the varying needs of people globally, a set of designs will offer a variety of features useful to different objectives and settings. CAPS have costs in constructing and maintaining the growing systems as well as harvest and component extraction.

Table 5.2 CAPS Design Parameters

Capability	Description
Hold, recycle and protect the water culture	• Provide a stable environment for growing water-based plants. • Recycle water to minimize fertilization costs and avoid ecological damage or pollution. • Minimize evaporation and water loss. • Avoid contamination from weedy opportunistic algae species. • Avoid contamination from bacteria, fungi, rotifers and other microorganisms as well as insects and birds.
Enable nutrient input and culture output	• Allow access for variable nutrient inputs. • Allow easy extraction of part or all of the culture or diversion to other containers. • Minimize evaporation and water loss. • Minimize contamination from opportunistic microorganisms.
Monitor	• Accurately measure important parameters. • Temperature, nutrient availability and pH. • Culture health, vitality and density. • Predators and weed algae.
Mixing	• Provide a means of mixing the culture to maximize solar exposure for each cell. • Keep cultures moving to minimize fouling. • Paddlewheel, moving blade or bubbling.

Design, production, maintenance, and costs are the focus of intense R&D to reduce the barriers to algae cultivation. Several companies, including NanoVoltaix, Inc. specialize in CAPS engineering. While many of the basic parameters are common, exactly how the CAPS are designed and operated vary based on the factors in Table 5.3.

Table 5.3 CAPS Design Determinates

Challenge	Design Drivers
What is to be produced?	• Food, food ingredients or health foods. • Fodder for animals, fowl or fish. • Energy for cooking fires. • Fuel for transportation. • Fertilizer and soil bioconditioner. • Water remediation. • Special chemicals or compounds. • Special micronutrients or nutraceuticals. • Medicines or vaccines.
Algae species selection	• Local – collect and test local species. • Distant – order from a credible source. • Test that species are robust and stable. • Insure a fit with the local environment. • Maximize to production or optimize production of one component. • Examine digestibility, taste, aroma, etc. • May grow single or multiple species.
Space with solar exposure	• Arid, desert or unused land. • Rooftops, balconies, sides of buildings. • Hillside, waste dump or empty lot. • Disturbed land such as old mines.
Materials available locally	• Acrylic, plastic, glass, tape and glue. • Clay, brick, cement or other useful materials. • Cloth or fabric for shading.

Inputs available locally	• Water – waste, brine, ocean or fresh.
	• CO_2 – density, bioavailability and cost.
	• Nutrients – waste water or other sources.
	• Labor – construction and operation.
Climate	• Solar exposure hours.
	• Cloudiness and degree of clouds and storms.
	• Daily temperature and temperature range.
	• Humidity, wind and weather extremes.
Local energy	• Foot pumps, bicycles, animals.
	• Solar or wind sources.
	• Electricity or small engines.
Cost	• Size – small, medium or large unit.
	• Minimize cost with local materials.
	• May use prefab with some local materials.
	• Nutrient, water and CO_2 availability.
	• Operated by a single person or family.
	• Operated by a community or co-op.

Vertical, angled or horizontal. Algae are solar collectors so the plants benefit from maximum exposure to the sun. Some angled CAPS track the sun similar to photovoltaic solar collectors. Horizontal CAPS, typically tubular or plastic bags, provide another variation in solar exposure. Several companies use plastic bags that are typically rectangular or oval.

Tubular CAPS

Different shapes provide different levels of solar exposure. A wide rectangle, similar to an aquarium, holds a lot of water but does not allow each alga cell to have sun exposure very often. Consequently, thin rectangular tanks, about three inches thick, tend to out-produce tanks that are wider. Tubular tanks may be a few inches wider because they present more surface area around the circumference. However, tubes around 6 inches typically out-produce wider tubes.

Experience with various designs in field settings shows the best architecture may vary by setting, Table 4.3. The algae biomass often cannot withstand ambient temperatures so they need protection. Without protection such as shading, cultures may be inconsistent in production, become unstable and simply stop growing. Fast growing pure algae cultures do not remain clean indefinitely and "weed" algae and predators must be removed.

Table 5.4 CAPS Type and Trade-offs

Type	Description	Limitations
Open pond	Economical, easy to manage, good for mass algae cultivation, considerable global experience.	Low culture control Stability issues Weak productivity High land and water use Species contamination
Vertical column	High mass transfer, good mixing with low shear stress, low energy consumption, scalable.	Small illumination surface Expensive construction Shear stress problems Cleaning issues.
Flat rectangle	Large illumination surface, good light path, good biomass productivity, relatively cheap, easy to clean, low oxygen build-up.	Scale-up challenges Culture stability Temperature stability Possible shear stress.
Tubular	Large illumination surface, good light path, relatively cheap.	Gradients of pH, dissolved oxygen and CO_2, tube fouling, high land use if laid flat.

The algae industry continues to experiment with variations in CAPS designs and operations. Most likely, low cost producers of commodities like feeds will use open systems. Applications producing algae oil for biofuels will probably use closed systems to maximize growth speed, vitality and species homogeneity. Maximum total production may be achieved with hybrid systems where CAPS grow pure strains quickly to desired densities, which then flow to open or covered ponds for production and harvest.

Added to the challenges of field settings is cost because in order to be commercially viable, algae production must occur at lower dollar and energy costs than energy alternatives. The National Renewable Energy Laboratory, (NREL), Algae species Program, for example, concluded in 1995 that closed systems were impractical for algae production because they were too expensive to build and maintain. Nearly all algae production to date occurs in open ponds but this will change quickly since most the planned production of algae for energy in the US will use closed or semi-closed production systems.

Abundance production models vary as much as traditional farms. Most experience with algaculture has occurred in ponds because they are cheap to build and relatively easy to operate. Ponds are most commonly built on a circular runway with a paddlewheel or other mechanism that keeps the culture moving. The turbulence between the surface and bottom continually mixes the culture and exposes all the algae cells to light.

Chapter 6. Entrepreneurial Opportunities

Opportunity is missed by most people because it is dressed in overalls and looks like work.

− Thomas Alva Edison

Abundant agriculture represents a new industry overlaid on the mature agribusiness industry that supports modern industrial agriculture. Modern farmers who want to achieve the abundance benefits can begin recycling their farm waste stream incrementally. Their initial successes will create huge demand throughout the abundance supply chain. The practice of abundance will quickly become as common as recycling domestic trash.

New industries create new and engaging entrepreneurial opportunities and jobs. Entrepreneurs define market needs and design, develop, and diffuse solutions that make businesses more effective and provide for customer needs. The new industry supporting abundant agriculture with inputs, farming, harvesting, and coproduct marketing creates a broad range of needed solutions. Initially, the dominant industry need is growing systems.

CAPS. Farmers need access to CAPS that are reliable, productive, and relatively inexpensive to construct and operate. Commercial models are available today but are typically 10 times too expensive for most farmers. The few commercial algae equipment companies today have focused on moderate to large scale systems where they can get a quick payback from their R&D.

Social entrepreneurs have a great opportunity to design, develop, and demonstrate small but scalable CAPS as illustrated in the prior chapter and in *Green Solar Gardens: Algae's Promise to end Hunger* (Edwards, 2009).

Unfortunately, much of the knowledge base about CAPS has been held confidential by companies as intellectual property. As a consequence, many costly and often fatal mistakes for business are repeated. Restricted IP means only few firms receive the advantages of technological, process, and CAPS breakthroughs.

The International Algae Competition – Visionary designs for sustainable algae food and energy at www.AlgaeCompetition.com plans to change current practice. The site orchestrated by Robert Henrikson, ex-CEO of Earthrise Nutraceuticals and the author is committed to open source knowledge and technology posted for global sharing, upgrading and use. The vision: "Distributed algae production systems globally will enable anyone on Earth with desire to produce sustainable and affordable food and energy locally for the needs of their family and community. We can collaborate to supplement the health and nutrition of our children while we advance social equity for all with community scale algae production systems."

The algae competition includes three tracks:

1. **Algae Landscape Design.** Design integrated CAPS into future landscapes, infrastructure and buildings.

2. **Cultivated Algae Production Systems (CAPS)** Develop working models and designs for CAPS and microfarms.

3. **Algae Food Development.** Create menus, new foods and food products incorporating algae as a featured ingredient.

The International Algae Competition engages people in all disciplines who will work independently, in teams and in concert to address the nontrivial challenges of building and operating cultivated algae production systems.

Remote monitoring technologies will revolutionize algae production. Today, each algae farmer must invest in a laboratory and learn biotechnology or hire scientists to monitor and maintain the algae culture. The process of selecting algae species, growing the initial cultures to sufficient densities to place in ponds, troughs or CAPS requires someone with considerable training. Failure to monitor the culture for health, weed algae invasion or predators can cause the entire production system to crash.

Remote Monitoring Equipment at the University of Arizona

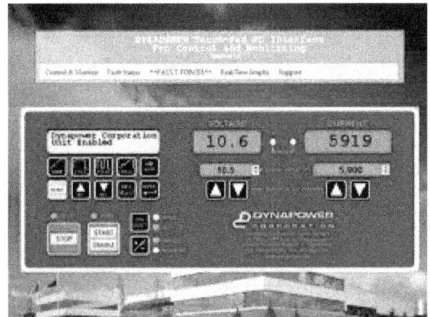 

Remote monitoring services will enable a single centralized laboratory staffed with scientists knowledgeable in biotechnology to support 100+ abundance farmers. Farmers could use an app on their iPhones with a digital microscopic camera to upload pictures of their culture several times a day. Some algae production facilities such as the University of Arizona have already set up relatively inexpensive 24/7 remote monitoring systems which can be observed at: www.xx

Harvest and extraction methods and equipment need to be simplified and cost reduced. New technologies show promise for low-cost harvest including new filtering methods, flocculation approaches, and ultrasonic sound stimulation that bursts the algae cells.

Growing algae for fish or plants, aquaculture, or hydroponics, eliminates the need for harvest because part of the culture simply flows into separate containers. Similarly, Smartcultures simply grows and flows the algae culture to fields in irrigation water, or by spraying.

Algae Harvest – Bucket and Sieve and Robot Harvester

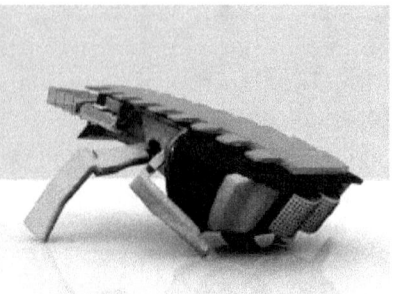

Nutrient recovery services will also be helpful to the industry. While some farmers will find nutrient recovery easy, others may need help in providing sufficient nutrients to meet their algae production goals. Buying the full set of chemical nutrients from commercial sources is not sustainable and too expensive.

Municipal and industrial wastewater treatment facilities represent a gold mine of recyclable nutrients. Currently, most these nutrients are locked in sludge, which is then buried or burned. New technologies to efficiently extract those precious nutrients would help the close the nutrient cycle and diminish pollution while providing cheap nutrients for abundance farmers.

Similarly, inexpensive extraction methods for CO_2 will be critical for abundance. Every ton of algae consumes two tons of CO_2, which is costly to buy at industrial gas prices.

Cars produce about a pound of CO_2 per mile. When an inventor creates a carbon capture filter for vehicle exhaust pipes, there will be a nearly limitless supply of low-cost CO_2 for growing algae. The filter would have to be removable because each 100 miles traveled would add 100 pounds to the vehicle's weight, and reduce gas mileage. Green consumers would put up with the extra hassle if the process were safe, easy, and fast. The filters could be exchanged and aggregated at service stations and used to feed algae production.

Small pyrolysis units will enable abundance farmers to burn their organic wastes in order to produce H_2 that powers an electric generator and CO that feeds the algae. Pyrolysis vents no gases to the atmosphere and creates a byproduct, biochar which is an excellent slow release fertilizer. Fortunately, several lines of research and practice are converging on the pyrolysis process for waste management and recovery. Breakthroughs in pyrolysis will enable manufacturers to offer small and medium scale units.

Pyrolysis Units

Light management innovations will be critical for farmers living in the northern latitudes. New techniques, such as bent light systems, fiber-optic cables, highly efficient LEDs and other methods will enable extended production where lights are an issue. Additional light modification innovations may be useful in the tropics where cultures may get too much bright light, which shocks the culture and causes growth to slow.

Culture management innovations will make growing algae more robust as farmers learn how to better measure and manage culture parameters such as temperature, pH, cell density, mixing speed, shear stress, contaminants, and a host of other issues.

Less than half the focus for abundance innovation needs to be technical solutions because the social and political issues are probably more important.

Social and political drivers

A variety of economic, social and health factors will accelerate abundance. The rising cost of fossil fuels and foods, while bad for consumers, will prompt more R&D and heighten consumer interest in SAFE production. As scientists and journalists, create a clear link between crop subsidies that create monocultures and the resulting health impacts on consumers, Congress will be motivated to shift subsidies to sustainable, ecologically positive and healthy production.

The Sierra Club, Worldwatch, Greenpeace, Audubon Society, World Future Society and many other environmental and social organizations are promoting green initiatives that help consumers understand the value proposition for SAFE production. Scientific organizations such as the Union of Concerned Scientists, the Environmental Working Group and The National Energy Independence Plan, NEIP, are providing the foundation for change in national energy policy. Strong voices including Al Gore, T. Boone Pickens, Bill Bailey, NEI, Ken Cook, ERG and many others help move public opinion towards sustainable and renewable energy solutions.

Some communities are orchestrating the systematic collection of used restaurant cooking oils and are home-brewing biodiesel. Arizona's Desert Biofuels Initiative's "Gold to Green" project, led by Brad Biddle, Sam West and the myself hopes to refine every drop of used restaurant cooking oil in Arizona to green diesel and remove 100 tons of fossil fuel pollutants from Arizona's air each year.

Sustainable communities with home-grown food, feed, and fuels may become as widespread as Victory Gardens during World War II. Individuals and cooperatives may build CAPS on rooftops, balconies, or backyards motivated by the desire to go green and to save money, eat healthy, nutritious foods, and save the planet. Local cooperatives might operate to process the green biomass into usable food and coproducts.

Existing social groups might assist people to build hybrid solar, wind and algaculture systems that use 100% off-grid, renewable energy. Algae production requires small electric motors (or human labor) for mixing water, supplying nutrients, monitoring, and extraction. These energy needs could be supplied by hybrid systems that use the sun to produce electricity. Solar power suits algae well since solar energy is only collected during the day and algae grow only when the sun shines. Tapping geothermal sources for warming greenhouses will enable year-round production in cold climates.

Abundance Designs for Cities

Sustainable communities may operate collectively to produce possibly 50% of the food and biofuels for their community. Communities may compete for bragging rights for self-sufficiency which means CAPS will sprout from vacant lots and rooftops. Community members may form cooperatives that specialize in specific foods, fuels or other coproducts – especially wine and beer.

Wine and beer made from algae grown in backyards or balconies will spur the explosion of abundance. People like a affordable good food but they will embrace with even more excitement the opportunity to start their own winery or beer garden. When the proper strains of algae are identified for wine and beer, people well beyond moonshine country will quickly jump on the home brew bandwagon.

Home brewing with grapes or hops is expensive and cumbersome. Typically, the grape juice feedstock must be purchased as well as the sugar, yeast, and other supplies and equipment. Individuals or cooperatives who homebrew algae wine or beer will find they can grow their own feedstock. Companies that sell hot tubs and hydroponic supplies are likely to gain a huge new source of business.

Home brewers will have to come up with a better name than algae wine since the term algae lacks cachet. Brewers will also have fun describing the wine's taste since classical features like "Earthy tones" may not be appropriate.

Several years ago, I proposed a name for the food component of the algae biomass after harvest – Alnuts. Consumers who have an aversion to algae may find Alnuts a more palatable name for their food, especially if it is formulated to look and taste like traditional foods. Alnuts combines the advantages of great taste, texture and mouth appeal with high protein, low calorie foods. Alnuts formulated as cake would provide great taste, be filling and be low in fat and calories. Alnuts could be made into texturized vegetable protein and substituted for soy or other vegetable sources. Algae foods present many opportunities for entrepreneurial innovation.

Snack foods will probably be the first algae-based foods that replace high caloric, low nutrient foods such as potato chips, snacks, French fries, and candy bars. High fat foods can be delivered with equal taste appeal but save on fats and calories. Algae doughnuts, breads, and pastries will provide excellent taste and texture as well as nutrition.

Low Calorie, High Nutrient Algae Foods

Fast food restaurants will offer healthier, tastier, texturized algae protein burgers, chicken, tuna, salads, and sushi. Consumers will have a choice; they can buy a conventional beef-based hamburger loaded with saturated fats for $9 or an algae-based texturized vegetable protein burger for $3 that tastes great, is just as filling, and provides substantial health benefits.

Health and energy bars and drinks will became very popular as food processors integrate high-value algae nutrients, vitamins, minerals, trace elements, and antioxidants into snacks and drinks.

Consumers may prefer texturized algae protein to another new meat source they will see on the shelves at about the same time; lab-grown tissue-cultured meats. The concept of lab grown meat is not a new concept. Winston Churchill in his 1932 essay *Fifty Years Hence* wrote: "We shall escape the absurdity of growing a whole chicken in order to eat the breast or wing, by growing these parts separately under a suitable medium." Scientists are growing tissue cultured meat in sheets that look similar to sheets of Nori, the popular seaweed used to wrap sushi rolls. The proposed name: sheats.

Premier foods made from edible species such Nostoc form small odorless, tasteless balls that appear similar to caviar. These might be processed with pigments and taste to imitate caviar. The low price point would make this form of caviar very attractive for consumers. A host of other premium foods such as truffles, saffron, or foiegras would also be popular.

Gourmet algae foods will also develop. Restaurants will differentiate themselves by serving gourmet algae-based foods. Popular cooking magazines will showcase algae foods on their covers and give prizes for best recipes and food innovations. Hybrid gourmet and health magazines such as *Cooking Light* will be the big winners because they can showcase phenomenal foods that are low in fat, cholesterol, and calories but high on visual impact, aroma, nutrition, texture, and taste.

The value proposition for abundance foods will include organically grown locally using minimal fossil resources. Abundance foods will be in low price with superior nutrients, health, taste, texture, aroma, as well as sustainability and ecological benefits. Institutional foods such as meals for schools, institutions, prisons, businesses and cafeterias will be able to provide healthier foods that taste better, offer better nutrition, higher protein, and less fat at a substantial price reduction over traditional foods.

Abundance foods will rise on the incoming tide of algae health initiatives that provide high-value innovation.

Game changers

Breakthroughs to any of the production obstacles will advance abundance and 360 algae microfarms. Green masterminds will find ways to lower costs and simplify production technology. Innovations that will accelerate adoption include the following.

Success models. The major impediment to abundance comes from the lack of successful models. When modern farmers are able to co-locate reliable and affordable 360 microfarms to support their existing farms, diffusion will move quickly through global collaboration.

Lasagna gardens. When families or communities can grow biomass effectively on the top of their 360 microfarm, fish in the middle and shellfish on the bottom, many people will become algae and fish farmers.

Mixed growth gardens. Moving the lasagna garden another step, mixed-use gardens add aquaculture to produce vegetables, grains and fruit, in the recycled water.

ZooPoo. Demonstrating abundance in zoos to recover, recycle and reuse the zoo animal and botanical waste along with the zoo trash would educate millions of visitors to the importance of waste stream energy and nutrient recovery. ZooPoo would provide considerable feed for a wide variety of zoo animals and fish. Demonstration facilities would also be great for aquariums, botanical gardens and science centers.

Prisons. The US currently has more than twice as prisoners as farmers. Prisons cost more per year than a Harvard education, consume extensive resources and create huge waste streams. Abundance could engage trustees in green collar jobs, remediate waste streams, produce algae biomass for tilapia and other fish, as well as organic fertilizer for local fields.

Famine, flood and disaster relief. When political leaders realize the cost of growing and shipping food is unsustainable, they will authorize funding for 360 microfarms that enable people in-country to grow their own food and coproducts. The same logic holds for foreign aid.

Special needs farms. Research on special needs kids and older people with dementia and Alzheimer's provides compelling evidence of vitality, health, and cognitive improvement that occurs from animal and plant husbandry. Special needs farms could integrate 360 microfarms into their food production as well as their energy and sustainability educational systems.

Biodegradable plastic. When 360 microfarms produce biodegradable algae plastics, the CAPS in which the biomass grows can produced by algae. An integrated production system might employ a few 360 microfarms to mass produce plastic that serve hundreds of other microfarms. Similar mass production of component parts, microscopes, monitors, mixers and other items necessary for algae production will be necessary to bring construction and operating costs to practical levels.

Toxin control. Finding cheap, low-tech, yet effective ways to control toxins and other microorganisms will be a tremendous benefit. Algae and algae-fungi monitors can provide low-cost solutions. Some microfarmers may use algae to remove toxins and pollutants from water, air and soil.

Clean water strategy. When a country such as Zimbabwe recognizes that 360 microfarms can remove the the bacterium *Vibrio cholera* that causes cholera and parasites that cause a host of other illnesses from water supplies, they will promote the adoption of abundance, assuming they can find funding.

Vaccine or medicine strategy. When a country plagued by malaria, SARS or HIV/AIDS decides 360 microfarms can provide cheap, yet effective medicines, roll-out will occur quickly.

Most likely, several excellent breakthroughs have already occurred, and yet no one has thought to apply the new technology to abundance production. An abundance collaboratory that systematically searches for novel solutions, while recognizing and rewarding innovators, will advance demonstration and diffusion. The collaboratory will also convey social entrepreneurial opportunities in many fields, especially food, energy, cosmetics, and health. We are building the collaboratory at www.AlgaeCompetition.com.

Algae cosmeceuticals

Both indigenous and modern people have used algae for thousands of years for food, fodder, fertilizers, vitamins, minerals, medicines, and cosmetics. Indigenous people used algae as we do today for natural treatment of skin moisture, bruises, burns, bites, stings, cuts, wounds, joint pain, headaches, and indigestion.

Seaweed and Microalgae

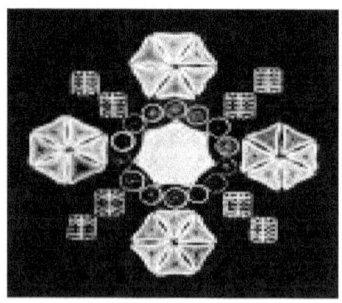

Algae are miniature biofactories that produce an extremely rich source of oils, carbohydrates, proteins, enzymes, fiber, and building blocks for food, cosmetics and medicines. Algae produce photosynthetic pigments in order to capture solar energy in chlorophylls, carotenoids (carotenes and xanthophylls), beta-carotene, lutein and phycobilins. Each pigment absorbs a different set of angstroms in the color spectrum.

Salmon, lobsters, crab, shrimp, and other shellfish get their beautiful reddish colors from the algae carotenoid astaxanthin.

Use of algae in lichens for pigments and dyes pre-dates Julius Caesar. The classic red color of Roman war tunics came from pigments extracted from the lichen urchilles. Roman women valued the plant and used it as rouge to give their faces a sensual color. Roman officers fed their horses algae to keep them healthy and to make give their coats a colorful sheen.

Algal oils and pigments are used today as cosmetics and skin moisturizers, similar to the use of aloe and jojoba oil. Algae building blocks form compounds that benefit the skin in many ways, Table 6.1.

Table 6.1. Examples of Algae Cosmeceutical Compounds

Group	Description
Emulsifiers	Many cosmetic products are based on emulsions. Emulsions are tiny droplets of oil dispersed in water or small droplets of water dispersed in oil. Since oil and water do not mix, emulsifiers produce small droplets and prevent the oil and water phases from separating. Emulsifiers work by changing the surface between the water and the oil, producing a homogeneous product with an even texture.
Preservatives	Preservatives are added to cosmetics to prevent the growth of microorganisms (e.g., bacteria, yeast, virus and fungi), which can spoil the product and possibly harm the user.
Thickeners	Thickening agents such as polymers are often added to cosmetics to change their consistency. Polymers can be synthetic (e.g. polyethylene glycol) or derived from natural sources (e.g. polysaccharides). Seaweeds are a natural source of polysaccharides. Carrageenans are extracted from red algae and alginates from brown algae. Cosmetics that are too thick can be diluted with solvents such as water or alcohol.
Fragrances, colors and pH stabilizers	Cosmetics typically include chemicals that give a pleasant aroma to the product, provide an appealing color, taste or texture or adjust the pH

(acidity). Customers prefer cosmetics with good taste and aromas.

Moisturizers

Moisturizers repair dry, burned or scaly skin. Skin becomes dry when water is lost from the top layer of dead skin cells faster than moisture can enter it from the living layers of skin below. Moisturizers can correct this problem in two ways: by preventing further moisture loss (occlusion)) and by adding substances that increase the water-holding capacity of the skin (humectants).

Shampoos and soaps

Shampoos and soaps clean by the use of surfactants – surface active agents. Surfactant molecules have both fat soluble (lipophilic) and water-soluble (hydrophilic) parts. The lipophilic part of the molecule sticks to oil and dirt, and the hydrophilic part allows water to then carry away the otherwise water-insoluble grime.

Lipstick

Water solubility - or the lack thereof - is an important factor in creating lipstick. Lipsticks are made by combining a water-insoluble dye with wax and a non-volatile oil such as beeswax. The resulting substance is firm but spread easily. Because it is water-insoluble, the lipstick will not be dissolved by saliva or drinking. Lipsticks often use dyes which react with the amino acids in the skin protein and change color.

Imitation tans

Fake tans also change color on contact with skin. The active ingredient in most fake tans is dihydroxyacetone, a colorless compound that darkens when it reacts with the amino acids in the top layer of skin. The color change is permanent but because skin cells are constantly being shed, the tan is usually gone after about a week.

About 90% of modern cosmetics contain algae extracts including agar, carrageenans, alginate and astaxanthin. Agar is mainly used as a preservative for meat and fish and as a gelling agent in food. Carrageenans are used in colorings, for cosmetics, toothpastes, ice-creams, pet foods, lotions and as a stabilizing agent in dairy products. Brown algae (kelp) is a source of alginic acid, which is used as a thickening, stabilizing and emulsifying agent in lotions, skin creams, ice-creams, dairy products, rubber, paint, shaving creams, adhesives and in the textile industry.

Kelp and Ulva lactuca

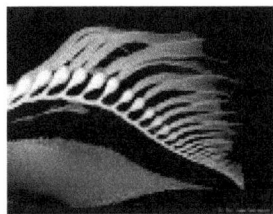

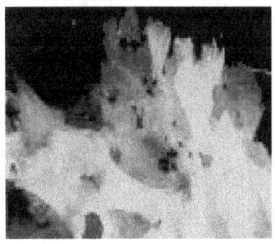

Algae commonly found in cosmetics include kelp, ulva lactuca, ascophyllum, Laminaria longicruris, laminaria saccharine, laminaria digitata, alaria esculenta, porphyra, chondrus crispus, and mastocarpus stellatus.

Currently, producers harvest many tons of seaweeds from natural stands to produce cosmetics and many coproducts. The value proposition for algal cultivation can be seen in the production of astaxanthin and other pigments that provide color to fish, animals and foods. Algae products also improve the sheen and body in hair.

Cultivation produced a natural hybrid (not genetically engineered) strain that produced 100 times more astaxanthin than wild seaweeds and was 10 times easier to harvest and extract. The cultivar also has five times the purity of naturally extracted astaxanthin.

Some firms such as Algae Biosciences have designed an algal biofactory production system that can target any useful compound, domesticate wild algal strains and train them to produce large amounts of the target compounds. Biotechnology research announces valuable new advanced compounds almost daily that improve skin, scalp, face and other cosmetic needs. These biofactories will be able to quickly adapt naturally occurring algal species to produce large amounts of the desired compounds that are extremely pure. Purity is critical for many cosmeceutical products.

CAPS can produce organic cosmetics with advanced compounds that smooth, protect, heal and promote skin regeneration. Algae produce a wide range of antioxidants which are a valuable ingredient in a spectrum of anti-aging formulations. No empirical evidence exists yet for any plant or synthetic source to remove wrinkles, but several algae compounds are known to play a major role in skin restoration and these can be produced. As new advanced cosmeceutical compounds are discovered, specialized biofactories can produce them quickly and economically.

With the exception of agriculture, the most engaging entrepreneurial opportunities for abundance may be health; especially public health.

Health initiatives

Today the four most prevalent deficiency diseases globally are malnutrition, nutritional anemia (iron and B12 deficiency), exophthalmia (vitamin A deficiency), and endemic goiter (iodine deficiency). Each of these nutrient deficiencies creates tremendous pain and suffering as well as severe economic and social drag. Nutritional deficiencies diminish the health, vitality and mental ability of those suffering. Nutrient deficient children often grow slowly, stunted, weak, lack energy, and are unable to think or learn normally. Nutrient deficient mothers endow their children with a heritage of mental retardation and physical maladies that are often not remediatable after birth.

Malnutrition. The UN World Health Organization, (WHO), recognizes malnutrition as the gravest single threat to the world's public health. Malnutrition plagues communities where families have insufficient access due to cost or availability of good protein or essential nutrients. Abundance production can supply sufficient affordable protein and nearly all the essential nutrients for normal development.

Anemia. Iron deficiency diminishes the health and vitality of over half the world's population, 3.5 billion people. Anemia is the most common blood disorder and creates a decrease in normal number of red blood cells or less than the normal quantity of hemoglobin in the blood. Iron is essential for strong red blood cells and a healthy immune system. Since human cells depend on oxygen for survival, varying degrees of anemia can have severe medical consequences. Anemia causes weakness, fatigue, general malaise, and brain dysfunction. Anemic children have trouble concentrating and learning. Severe anemia can cause loss of breath and cardiac arrest. Anemia typically occurs from insufficient iron in the diet.

Algae provide the richest source of B-12, 10 times higher than high iron foods like beef liver. Vitamin B-12 is necessary for development of red blood cells, especially in the bone marrow and nervous system. One tablespoon of algae a day or the meat of algae eaters such as fish eliminates iron anemia by providing sufficient folic acid (vitamin B-12). Algal iron is easily absorbed by the human body because its blue pigment, phycocyanin, forms soluble complexes with iron and other minerals during digestion making iron more bioavailable. Algal iron is over twice as absorbable as the form of iron found in leafy vegetables and meats.

Vitamin A deficiency. Excellent vision, especially night and color vision, requires substantial amounts of vitamin A, yet nearly half the children in the world today are deficient. Vitamin A is important in maintaining mucous membranes and pigments necessary for vision. is The eye depends on Vitamin A in the retina to form of a specific metabolite, the light-absorbing molecule retinal. This molecule plays a critical role in both night and color vision. Vitamin A deficient lactating mothers pass the deficiency to the child, which may disrupt embryonic development. Vitamin A also plays an important role in other human systems including gene transcription, immune and cardiac function, bone metabolism, hematopoiesis, skin health, and antioxidant activity.

The most common cause of blindness in developing countries today occurs from vitamin A deficiency. The WHO estimates 13.8 million children have some degree of visual loss related to vitamin A deficiency. Approximately 500,000 children in the developing world go blind each year from insufficient vitamin A and nearly half of those children die within a year of becoming blind. Night blindness is a marker of vitamin A deficiency, which can also lead to impaired immune function, cancer, birth defects, and maternal mortality.

The Kanembu tribe in Chad avoids vitamin A deficiency, using a strategy they have used for centuries, by adding a few grams of locally harvested algae to their meals each day. Various algal varieties provide ten times the beta-carotene (a provitamin A carotenoid) per pound than carrots. Vitamin A deficiency, often accompanied by zinc and manganese deficiency, amplifies the health impacts. A single tablespoon algae supplement provides sufficient daily zinc and manganese for adults and children.

Field studies in several countries have demonstrated that two grams of algae a day can cure vision impairment in children and adults. Algae are also rich in phytonutrients and functional nutrients that activate digestive and immune systems.

Iodine deficiency. According to WHO, in 2007 nearly 2 billion people have insufficient iodine intake, making iodine deficiency the largest preventable cause of mental retardation. Even moderate iodine deficiency in pregnant women and infants, lowers intelligence by 10 to 15 I.Q. points. The most visible and severe effects include disabling goiters, cretinism, and dwarfism. About 16% of the world's people today have at least mild goiter, a swollen thyroid gland in the neck. The high iodine content in algae contributes to the low rates of goiter observed in countries where people frequently eat algae.

It may seem improbably that a tiny algal supplement can provide sufficient iodine, iron, zinc, and other nutrients even when the local diet does not. Typically, these critical trace elements exist in the local water but in extremely weak dilution. People, especially children, are unable to drink enough water to acquire sufficient iodine. In many cases, little fresh water is available for drinking. Algae's secret to high nutrient value stems from its ability to bioaccumulate nutrients in water at 1,000 ambient levels. This means that even when some nutrients, minerals or vitamins may be lacking in human diets, algae can concentrate those nutrients in the green biomass.

Imagine how valuable globally distributed abundance production will be for the billions of children and adults that suffer from hunger, malnutrition and nutrient deficiencies. Local algae production will give these people health and a hope for a better future.

Algae provide even more benefits for plant health than human health. That is why smartcultures are so effective in stimulating crop growth and improving yields.

Chapter 7. Soils and Fertilizer

The more we pour big machines, fuel, pesticides, herbicides, fertilizer and chemicals into farming, the more we knock out the mechanism that made it all work in the first place.   **– David R. Brower**

Fertile soil is not an inert medium but a mixture of water, air, minerals and organic matter. In most soils, minerals represent around 45% of the total volume, water and air about 25% each, and organic matter 2-5%. The mineral portion consists of three distinct particle sizes classified as sand, silt or clay.

Soil health depends on the organic component that houses many living creatures along with dead material in various stages of decomposition. An acre of living soil may contain 900 pounds of earthworms, 2400 pounds of fungi, 1500 pounds of bacteria, 133 pounds of protozoa, 890 pounds of arthropods, algae, and possibly some small mammals. An acre of soil may contain over 10,000 species of microorganisms which contributes significantly to the biodiversity inorganic soil. Appendix I provides a brief description of the primary soil macro and microorganisms.

_____

Note: This chapter is adapted from *Smartcultures: Sustainable Food despite Climate Change and the Mass Extinction of Fossil Resources*, Chapter 9, Mark Edwards, 2010.

Soil organic matter interacts to influence soil biological, chemical, and physical properties and consists of:

- Raw plant residues and microorganisms (1-10%)
- Active organic traction (10-40%)
- Resistant or stable organic matter (40-60%) also called humus.

People talk of agricultural practices that are differentiated by their input source – chemical (synthetic/industrial) or organic. From a plant's point of view, synthetic versus organic has no meaning because plants are blind to source. If a plant's roots find a needed bioavailable nutrient, plants absorb the nutrient and put it to use building structure or storing energy. What plants care about and the factor that makes all the difference in produce quality is the "fountain of energy flowing through the circuit of soils, plants, and animals; fertility." Soil fertility depends on maintaining sufficient nutrients, minerals, organic matter, moisture, and microorganism communities.

The magic of natural soil fertility comes from the combined work of earthworms, arthropods, and the microbial community, which Nathaniel Hawthorne recognized when he observed: "The divine chemistry works in the subsoil."

Crop harvest and erosion work in tandem to remove nutrients from soils, so they must be replaced with fertilizers. Fertilizer is the most widely used material in agriculture and comes in three types: chemical, organic, and biofertilizer. Chemical fertilizers are also called mineral, synthetic, and inorganic.

Fertilizers are usually applied directly onto the soil, but can also be applied onto leaves, trunk, or exposed roots (foliar feeding). Fertilizers can be either organic (e.g. manure or compost) or inorganic (mined or synthesized chemically). Organic fertilizers have been used for centuries whereas chemically synthesized inorganic fertilizers were developed in the 20th century. Partial fertilization also can occur through biological processes like N_2 fixation or biofertilizers. Globally, mineral fertilizer is the major pathway for nutrient replacement because synthetic and mined fertilizers have been cheap, available, and relatively easy to apply.

Experiments with added N doubled grain production but the addition of N and P together produced five and six-fold increases. Governments intervened directly to assure the supply of fertilizer to farmers through policies of price-fixing and subsidies. By the mid-1980s, subsidies for fertilizers reached 70% of the world price, pesticides 40%, and water 90%.

Chemical, organic and biofertilizers

The advantages of each fertilizer type need to be integrated in order to achieve optimum crop growth, Table 7.1.

Table 7.1. Chemical, Organic and Biofertilizers

Advantages	Disadvantages
Chemical Fertilizers	
Typically cheaper and more available than organic fertilizers.	Diminishes produce taste and texture, shortens shelf life, decreases produce quality and reduces the total digestible nutrients.
Some nutrients are soluble and immediately available to the plants. The effect occurs very quickly.	Oversupply of N softens plant tissue resulting in crops that are more sensitive to disease, weeds and pests. Watery tissues also negatively affect taste and texture.
High in nutrient content so only modest amounts are required for crop growth.	Over-application results in negative effects such as leaching, water pollution, destruction of microorganisms, and friendly insects, crop susceptibility to diseases, acidification or alkalization, and reduction in soil fertility.

Light and far easier to apply than organics.	Adds toxic metals and minerals to produce.
	Reduces the colonization of plant roots with microbes. Inhibits natural symbiotic N fixation by rhizobia (soil bacteria) due to high N fertilization.
	Accelerates the decomposition of organic matter which degrades soil structure and causes erosion.
	Nutrients are easily lost from soils through binding, leaching or gas emission, which diminishes fertilizer efficiency.

Organic fertilizers

Nutrient supply is more balanced which makes plants stronger and healthier.	Comparatively low in nutrients so larger volumes are needed to provide sufficient nutrients for crop growth.
Enhances soil biological activity, which improves nutrient mobilization from organic and chemical sources and decomposition of toxic substances.	Heavy to transport and apply. Transportation costs may make use impractical. Nutrient release rate may be too slow to meet crop growth requirements.
Enhances the colonization of mycorrhizae, which improves P supply.	Major plant nutrients may not exist in sufficient quantity to sustain maximum crop growth.

Enhances root growth due to better soil structure and encourage the growth of beneficial microorganisms.	Nutrient composition of compost is highly variable and in developed countries, the cost is high compared to chemical fertilizers.
Increases soil organics, improving the exchange capacity of nutrients, improving water retention and buffering the soil against acidity, alkalinity, salinity, pesticides, and toxic heavy metals.	Long-term or heavy application may result in salt, nutrient or heavy metal accumulation and may adversely affect plant growth, soil organisms, water quality, and animal and human health.
Releases nutrients slowly and contributes to the residual pool of organic nutrients in the soil, reducing N and P leaching loss; also recycle minerals and micronutrients.	Manure or compost must be plowed into the soil or the nitric oxides will volatize.
Helps to suppress certain plant diseases, soil borne diseases and parasites.	Animal manure may contain heavy metals and pharmaceuticals used in animal production.

Algal biofertilizers

All the advantages as organic fertilizers, plus:	May not be sufficiently effective on dry land farms. May need irrigation for nutrient delivery.
Improves soil porosity, cultivation cost, light and water penetration,	Displaces only up to: • 50% of chemical fertilizer • 30% of water use

moisture retention, seedling emergence, and efficient gas exchange.	• 30% of fossil fuel use • 80% of ag chemicals and poisons
Delivers full range of macro and micro nutrients.	May not be available locally yet.
Low cost delivery with irrigation or by sprayer.	Sparse research exists on growth stage specific nutrients.
Grows strong plants that have more resistant to heat, pest, disease, and drought stress.	Sparse field research except for rice and hydroponics.
Reduces need for: • Agricultural poisons. • Energy and fertilizers. • Cultivation	Novel and requires some change in farming methods.
Delivers special nutrients and hormones that stimulate plant growth.	Biofertilizer support industry does not yet exist.

Fertilizer cost structures are likely to change with increases in fossil fuel prices and the coming scarcity of mined inorganic minerals. Industrial fertilizers will become far more costly when communities begin to levy pollution and health taxes. Organic fertilizers will rise with the cost of fossil energy. Biofertilizer prices will decrease while they increase in nutrient value as we learn enhanced production methods, improved technologies for plant growth regulators, and more about novel microbial tools for improving soil structure.

Forms of Fertilizer

Algal biofertilizers

Algal biofertilizers (or inoculants) are a low cost, effective, environmental friendly and renewable source of plant nutrients that supplement, and in some cases replace chemical fertilizers. Algae are photosynthetic organisms with chlorophyll that enables them to produce organic matter by photosynthesis (CO_2 fixation) and in some species, N_2 fixation. Microalgae live in symbiosis with lichens and mosses and make up cryptogamic crusts, which are often the major sources of biologically fixed N_2. Soil crusts act as a protective covering to minimize topsoil erosion from water and wind. Crusts may also provide the soil structure necessary for seed germination.

Algae are ubiquitous members of soil microflora and offer numerous advantages as biofertilizers. Algae do not compete with crops or other soil microflora for carbon since algae captures carbon along with N_2 from the air. Algae do not compete with crops for energy because neither can absorb more than a small fraction of the available sunlight. Fixed N_2 and other nutrients in algae become bioavailable to crops as a combination of leached N from living filaments and mineralization of decaying algal biomass.

Algae stimulate production of natural plant growth hormones that accelerate cell division and elongation, producing taller, greener, and lusher plants that produce higher yields. Algae also stimulate plants to secrete compounds that repress harmful bacteria, fungi, and other pests. In some cases, algae operate as a catalyst that helps plants manufacture natural insect repellent on their leaves.

Good soil consists of about 94% mineral and 3-7% bio-organic substances. The bio-organic parts are 85% humus, 10% roots and 5% edaphon. Microflora decomposes organic materials to produce humus. Edaphon may be the smallest soil component but plays a critical role for plant growth and development. Edaphon consists of microbes, algae, fungi, bacteria, earthworms, microfauna, and macro fauna. Beneficial microorganisms operate to maintain the ecological balance by active participation in nature's carbon, N, sulfur, and P cycles. Soil microorganisms also play a pivotal role in building and enriching fertile soils. Algae also expand soils and increase porosity, which makes room for the colonization of soil microbes.

Physical, chemical and biological factors influence the growth of algae on and in soils. Parameters include light, temperature, day length, nutrients, salinity, pH, atmospheric humidity, desiccation, wind, predators, pests, density, and cycles of wetting, and drying. Some species are intolerant to too much light and go dormant. Many algal species thrive and fix N_2, optimally between 30° and 35°C, but are not active at higher or lower temperatures. In many cases, indigenous algae species can be found locally that have adapted to local light and temperature conditions and outperform pure laboratory strains.

Chemical properties are dominated by pH which influences species composition, growth and N_2 fixation. Algae grow best in neutral to alkaline soils. Nutrient and agrochemicals influence the activities in growth of algae and over application can be toxic. The addition of lime ($CaCO_3$) to fields stimulates algal growth.

Macroalgae biofertilizers

Most human societies that lived near coastlines harvested seaweed for human, animal, and plant food because of its excellent nutrient profile. Macroalgae typically have vital minerals, vitamins, and micronutrients such as iodine and vitamin A that are unavailable in sufficient quantities from local foods. Seaweeds act as natural fertilizing agents and stimulate helpful bacteria in the soil that release bound soil elements, fix N_2, and make nutrients available to the plants.

Considerable research has focused on biofertilizers derived from macroalgae (seaweed) such as kelp, Ascophylum Nodosum and fossilized kelp from calcium deposits, which contain a broad spectrum of trace minerals. Some species of macroalgae have a slightly acidic pH of 5.5, created by the presence of amino acids, which enables it to help balance alkaline soils.

Seaweed biofertilizer and soil conditioners are made up of an array of water soluble minerals. While chemical fertilizers typically have only NPK nutrients, biofertilizers may have 75 minerals, growth hormones, cytokinin, auxins, vitamins, and enzymes. Biofertilizers stimulate organic activity in the soil and lower toxic residues from various salts and chlorinated hydrocarbons. Activated plant growth hormones also reduce toxins from harmful organisms such as nematodes and fungi.

Biofertilizers decrease the need for insecticides by presenting stronger plants that take natural self-defense to insect invasions. For example, some fields treated with biofertilizers enable plants to produce a distasteful waxy film on their surfaces to repel insect attacks. Insects bypass treated fields in favor of untreated plants.

Algal biofertilizer improves the health and density of beneficial soil microbes which suppresses the proliferation of pathogenic microbes. Hardier plants build stronger cell walls due to the presence of additional silica (Si) which provides resistance to temperature spikes.

The research presented here used macroalgae or fossilized algae from calcium deposits harvested, processed, packaged, and shipped to growers in a manner similar to traditional fertilizers. We believe the smartcultures model, where the grower produces algae locally for addition to irrigation water, will produce equal or better results. The preliminary research has been very encouraging.

Only NDAs, non-disclosure agreements prevent empirical results from being presented here. The intellectual property issues hold the algae industry hostage. The descriptions that follow are based on empirical research on algae biofertilizers.

The positive growth characteristics from biofertilizers are only partially attributable to algae itself. Many of the growth benefits occur because algae attract a diversity of beneficial soil microbes and microfauna that work in symbiosis to support plants. The microbes stimulate natural growth compounds that lead to accelerated growth and hardier plants that are more tolerant to stressors.

Biofertilizer research

Most the research on macroalgae biofertilizers compared with chemical fertilizers has not been peer reviewed. Far more research on biofertilizers has occurred India and China than anywhere else. The literature in the US is sparse and from the 1970s.

The results presented here come from interviews with growers and suppliers of fossil biofertilizers that are mined fossil macroalgae. The nutrient material has quality variability but growing results appear consistent. Biofertilizer suppliers mine, package, and ship the product to retailers or growers. A full range of biofertilizers is available from hydroponics and urban garden stores. While fossil biofertilizers are no more sustainable than mined chemical fertilizers, the growing results are encouraging for locally grown biofertilizers.

Research on macroalgae biofertilizers compared with chemical fertilizers has not been peer reviewed. In Interviews with fertilizer producers, distributors and growers, industry members typically report significant (20% or higher) improvements in germination, plant density (due to higher germination rates), earlier maturation and larger, heavier yields with better taste. Biofertilizers reduce chemical fertilizer requirements and increase the plants' ability to withstand stresses from disease, insects, heat and drought. The results here are for mined macroalgae biofertilizer. Both macro and microalgae typically deliver the full set of macro and micronutrients plants need. The results here are for macroalgae biofertilizer.

Biofertilizer research

The positive growth characteristics from biofertilizers are only partially attributable to algae. Many of the growth benefits occur not from the algae but from the diversity of beneficial soil microbes and microfauna algae attract. These microbial communities work in symbiosis to support plants. The microbes stimulate natural growth compounds that lead to accelerate growth and produce hardier plants that are more tolerant to stressors.

Rice germination rates and growth rates improved, creating higher yields, earlier maturation, larger heads, higher protein content and lower fertilization requirements. Biofertilizers create improved germination rates for other grains, better stands, larger heads, and consistently higher protein content. The increased bioavailability of vital nutrients enables crops to mature earlier and to withstand drought conditions better than untreated plants.

Corn increased germination rates, ear size, kernel size and regularity, protein content, earlier maturation, increased yields both in silage crops and feed corn, increased ability to withstand disease and insect infestation and an increase in sugar content of the corn milk.

Hay grows faster and a leafier, faster recovery after cutting, lower water requirements, an increase in protein content and an increase in overall yields of 25% over the controls.

Cotton increased germination rates, grew faster, expressed more blossoms, more squares, and heavier setting of fruit. Plants displayed less loss dropping from the blossom, better boll setting, sturdier stems, heavier setting of seeds in the boll, increased luster to the fiber of the cotton itself, increasing the grade and the price. The crop had lower N and water requirements, better disease resistance, blossoms setting at the top of the plant while bolls were opening at the bottom of the plant, producing a longer fruiting period and higher yields.

Tomatoes grew faster with larger with an average of 10-14 days earlier maturing rate. Fruit was juicier, redder, with lower acid content, improved flavor, a significant resistance to disease (principally the mosaic virus), with increased yields of 10–23%.

Citrus trees grew faster, display a marked increase in sugar content of the fruit, thinner rinds, heavier fruits, higher disease resistance, lower fertilizer requirements, increased frost resistance (both to the tree and the fruit). Mineral deficiencies were far less prominent.

Fruit trees had heavier yields of all fruits tested (including peaches, pears, plums, apples, apricots, nectarines, and cherries), earlier maturation rates, heavier fruits, later fruits, better quality yields and lower fertilization requirements.

Sugar beets Displayed improved germination rates, faster growth, increased size and sugar content, more disease resistance and lower fertilization requirements.

Sugar cane Grew faster with earlier maturation, improved sugar content in quality as well as quantity and Higher yields.

Melons had higher germination rates, faster growth, higher sugar content, resistance to splitting and sunburning, earlier maturation, greater consistency in quality, lower water and fertilizer requirements, increased disease resistance and better quality retention (shelf life) after harvesting.

Soy beans improved germination rate by 22% germination rate over 32 different trials, 29% more nodulation in the rhizosphere, 21% yield increase, 9% protein increase, better disease resistance, lower requirements and earlier crop maturation.

Macroalgae biofertilizers offers significant value but have several downsides. Suppliers typically mine macroalgae from the sediments of ancient oceans or harvest seaweed along coastlines. The limited sources can supply only a tiny percentage of the fertilizer needed for industrial food production. Harvesting, packaging and transporting macroalgae consumes substantial amounts of energy. Macroalgae from marine environments often carry salts, which may degrade soils. Federal agencies currently do not regulate manufacturers, so growers cannot depend on the fertilizer quality.

Microalgae biofertilizers

Most the research on microalgae biofertilizers has been performed in India. Of the 400 million farmers in the world, approximately 100 million farm in India. India has few fertilizer mines and recognized 80 years ago that fertilizers and agricultural chemicals are nonrenewable resources. India's scientist began exploring fertilizer production that avoided resource depletion and environmental degradation. Scientists recognized the impending energy crisis due to fast depleting mineral oil reserves and begin a search for N_2 production using biological processes that used nature's mimicry *in lieu* of the energy intensive Haber-Bosch protocol.

India's soil scientists recognized algal biofertilizers as a perpetual source of nutrients that did not contaminate groundwater or deplete fossil resources. Various biological systems capable of fixing atmospheric N_2 combine to contribute about 175 million tons of N to plants every year as compared to about 50 tons being fixed by industry.

Biofertilizers are microorganisms which add, conserve, and mobilize the crop nutrients in the soil and can lead sustainable crop production. The primary microorganisms used as biofertilizers belong to bacteria, blue-green (cyanobacteria) and green algae.

Biological N fixation represents an inexpensive source of N for increasing the productivity of crops. The biomass of these organisms decompose rapidly in soils and supply significant amounts of N-P-K, sulfur, zinc, iron, molybdenum, and other micronutrients. The organic acids released during the biomass mineralization process accelerate P and micronutrient availability to crops. Biofertilizers amplify soil microbial populations such as bacteria, fungi, and other microflora that activate soil enzymes which improve soil fertility.

India coined the term "Algalization" during the 1960s to describe the biofertilization of rice soils, using free living blue-green algae. Algalization is a form of biofertilization where the fertilizer is grown in ponds, harvested and dried and then transferred to the field.

The government of India supplied a dried starter algal culture to Indian farmers who grew the algae for two months in small ponds. Dry algal flakes from the ponds, along with additional P and Fe were added to rice fields one week after transplanting rice seedlings. Algalization provided about half of the N needed for crop growth and development. Grain yields increased 33% with added inorganic N and about 16% with no added inorganic N.

Biofertilizers are generally applied to seeds, seedlings, or soils but are unlikely to have significant positive impact on plants unless they are able to grow and multiply. The introduced microorganisms will decline within days or weeks unless soil conditions support their growth. Inocula formulation and application method are critical for biofertilizers. Effective inoculation management includes concerns about shelf life, suitable carrier materials, susceptibility to high or low temperatures or humidity, problems in transportation, storage, and application.

Seed inoculation applies specific microbes that can grow in symbiotic association with plant roots. Soil conditions must be favorable for the inoculants to perform well. Selected strains of N-fixing rhizobium bacteria have proven to be effective as seed inoculants for legumes. Seed inoculation may occur with multiple types of microorganisms. Seed treatment with Rhizobium, Azotobacter or Azospirillum, coats the seeds and allows them to dry. Each seed has several layers of the material as the inoculants treat the outer seed layer. The layering process maximizes bacteria to generate better results.

Microbes added to the soil compete with microbes already living in the soil that are already adapted to local conditions. These indigenous microbes outnumber the inocula substanitally. Inoculants of mixed cultures of beneficial microorganisms have considerable potential for controlling the soil microbiological equilibrium and providing a more favorable environment for plant growth and protection. Joint inoculation of biofertilizer with mycorrhizae and N_2-fixing bacteria has been successful.

Biofertilizer limitations

In their review of biofertilizers, Wani and Lee concluded that dominate characteristic common to most biofertilizers in India lies in the unpredictability of their performance. Crop responses to biofertilizers are not as dramatic immediately as those with chemical fertilizers. These biological agents are subjected to a spectrum of hostile factors and their survival and efficiency is governed by variables such as the host plant, soil fertility, organics, and acidity. Other variables include soil moisture, cropping practices, biological and environmental factors. Biofertilizer effectiveness also depends on the presence favorable microorganisms and the absence of predators.

Barriers to biofertilizer adoption in India include poor quality of the inoculants produced, lack of knowledge about inoculation technology for extension personnel and farmers, effective inoculant delivery and supply systems, and lack of government policies to exploit biofertilizers. In India during the 70s and 80s, many small biofertilizer producers began selling poor quality inoculants to farmers who lost faith in biofertilizers. Once growers have lost faith in a growing method, it is doubly hard to recover their confidence.

Summary

Globally, farmers face an acute need to find cheaper sources of renewable plant nutrients. Sustainable food production requires fertilizer application that repairs ecosystems rather than damaging them. Many organic farming advocates frame their arguments that chemical fertilizers destroy soil structure and the application of organic fertilizers can recoup the loss of soil tilth. Others advocate a judicious combination of chemical and organic inputs to meet the shortfall of chemical inputs. Sustainable food production may need the appropriate combination of chemical fertilizers, organic manures, crop residues, composting, and biofertilizers that are most affordable.

The critical question is whether there is enough organic matter and microbial inputs available for intensive farming. The supply question has yet to be resolved but biofertilizers can play a critical role in both affordability and sustainability of the inputs needed for crop production. Algal biofertilizers that recycle farm wastes offer a cost-effective method for replacing extracted nutrients, rebuilding extracted or eroded organic matter, and enhancing soil structure.

Chapter 8. Food Production Alternatives

Every gun that's made, every warship launched, every rocket fired, signifies a theft from those who hunger and are not fed, those who are cold and not clothed. This world in arms...is spending the genius of its scientists, the sweat of its laborers

– Dwight David Eisenhower

Most of us hope that technology will once again rescue us from the vulgarities of climate change, resource depletion, and the rising cost of food and transportation. Based on population projections that anticipate our children will be sharing our planet with an additional 3 billion people, we will need every food productivity innovation available. The current foreign aid models are not sustainable.

Expand foreign aid

Countries globally have contributed money, technology, and food for people impacted by natural disasters on every food-growing continent. Global aid has served as a surrogate insurance policy to ensure food supplies. Unfortunately, global climate chaos will raise ocean surface temperatures and increase the number, frequency, and intensity of storms, temperature spikes, and droughts that cause natural disasters, Figure 7.1. Countries will lose patience with natural disasters and run out of discretionary money for foreign aid.

Figure 8.1 More Floods in Recent Decades

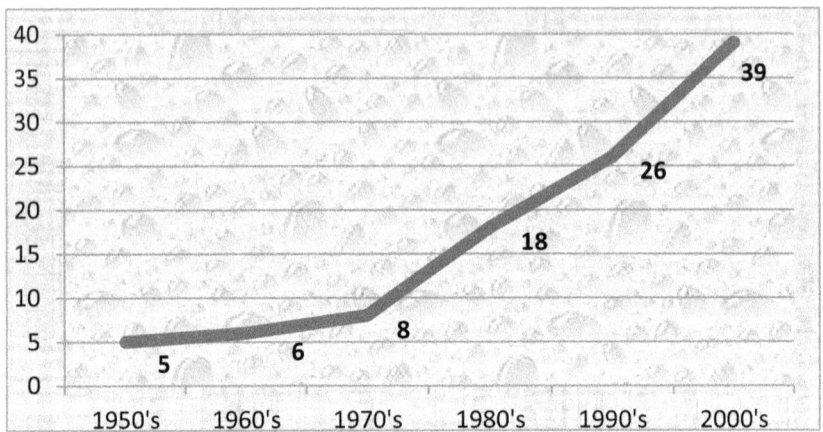

Global warming causes less rain to fall. The period between rains tends to be longer and rain falls not as farmers' rain but in violent storms. Downpours and floods multiply damage because intense rain causes water to run off rather than leaching salts. Violent storm run-off does treble ecological damage as it erodes thin topsoil, carries dissolved salts, and moves fertilizers and pesticides to waterways. The Union of Concerned Scientists predicts that 100 year floods are likely to occur once a decade.

The strategy of gifting food works only for the relatively few countries that currently enjoy good weather and have food surpluses. Gifting food is extremely expensive due to the added cost for storage, transportation and distribution. The UN Food and Agricultural Organization, (FAO) reported that shipping food aid can cost four times more than supplying the inputs for food production locally. The US has subsidized the substantial extra cost, which served as a hidden additional subsidy for American farmers.

A strategy for supplying food inputs has worked for decades but may not work in the future. Fertilizer price increases have caused India and other countries to cut or end fertilizer subsidies. Several countries in the Mid-East have stopped subsidizing irrigation water due to scarcity and expense. These countries now import their water in the form of food since one ton of grain consumes 1000 tons of water. As fossil resources become increasingly scarce and expensive, countries will have insufficient wealth to send the inputs necessary for growing food.

Expand industrial agriculture

Countries that are considering adopting modern agriculture practices should take a hard lesson from India's experience. India will overtake China as the largest population within the next decade and may have trouble supplying sufficient good food for 1.2 billion people. India was food independent for nearly a decade, partially due to the adoption of genetically engineered, (GE) food crops. Now, due to a combination of environmental and economic factors, amplified by GE crops, India must import millions of tons of grain to sustain its population.

Farmers in India

The US exported its "Green" Agricultural Revolution in the 1980s and sent money and technical support to India. This was a gift that many of India's farmers came to regret. India's government supported farmers with low-cost chemicals and new GE seeds in the 1990s. In the years since, India's farmers have discovered that the Green Revolution was not green in the sense of supporting sustainable food crops. Today, many farmers in India can afford neither the GE seeds nor the substantial fossil resources required to support GE crops.

A growing plant has only so much energy and divides that energy among competing components; including roots, shoots, and fruit. In their quest to create higher yields of heavier fruit, genetic engineers stole energy from the roots and shifted it to the fruit. The unintended consequence of this high yield design is that GE farmers must compensate for the diminished plant foundation, deep root system, with extra fossil resources. These additional inputs are not cheap and include fuel, fertilizer, fresh water, and cultivation as well as additional herbicides, pesticides, and fungicides. These fossil resources add substantial production expense, which forced many farmers out of business when they found the crop inputs unaffordable or unavailable. India's experience with GE crops illustrates catastrophic intended consequences.

Over 8 million Indian farmers quit farming during the 1990s due to rising GE crop input prices that escalated farmer debt. In the decade ending in 2007, 183,000 farmers in India committed suicide because their farms could no longer provide for their families. Government sources note that farmer suicides are substantially under-reported. Additional millions of farmers and family members have died become disabled from agricultural chemicals and poisons. Trains from the city of Chotia Khurd in northern India are now called cancer trains because so many people from the farming villages must go to the city for cancer treatments.

India's brutal lessons from modern agriculture include the following:

- **GE seed prices escalated** beyond the reach of many farmers. Farmers took out loans to pay for seeds but could not pay back the loans when their crops failed. They lost their farms.
- **Fertilizer subsidies escalated and then failed**. India's government recognized that farmers needed fertilizer subsidies for their high maintenance GE crops. Subsidies in 2005 totaled $4 billion but rising costs for fertilizer imports and diminishing supplies increased subsidy costs to $22 billion in 2008. The ballooning costs have prompted calls to reform the program that India depends on to maintain its food supply.
- **GE crops consumed substantially more fresh water.** Irrigation increased over 600% over the last two decades because high-yield GE seeds require substantially more water. The additional water improves seed germination, fertilizer absorption, and plant growth. The shallow roots of GE crops limit root zone depth, which necessitates additional irrigation when water percolates

below the root zone. Over a million wells have gone dry, ending the farmer's ability to grow crops.

- **Competing weeds developed resistance.** GE crops are designed for yield and cannot compete with natural weeds. Unfortunately, many invasive weeds have become resistant to Roundup™, defeating the transgenic advantage.
- **GE crops required substantially more cultivation.** GE crops are so vulnerable to natural weeds that farmers have to break up the topsoil before planting in order to remove competing weeds. Additional cultivation is required while the crop grows. The additional cultivation increases production cost and amplifies soil erosion as well as ecosystem pollution.
- **GE crops needed substantially more agricultural chemicals.** GE seeds are extremely vulnerable to insects, worms, fungi, mold, mildew and blight. Even though farmers applied tons of pesticides, insects and other pests have become resistant and typically destroy larger portions of each crop in India than they did two decades ago.

Fertilizers, pesticides, herbicides, fungicides, and other agricultural poisons have created havoc in India as monsoon rains moved massive amounts of soil and embedded chemicals into waterways. Monsoon rains flood fields and enable agricultural chemicals to migrate into surface and groundwater where they poison aquatic life, animals, and farm families.

India's experience with modern agriculture is not unique. Every farmer who plants transgenic crops knows that the fossil resource input costs will escalate – along with seed prices. Farmers also know that their fields are only sustainable as long as all the fossil inputs required for GE food production are affordable and available, precisely when they are needed. When the first fossil resource becomes too expensive or runs out, farmers have no choice but to abandon their fields.

Genetic engineering

In spite of India's experience, many people espouse GE seeds because genetic improvements formed the basis, along with expanded irrigation and chemical fertilizers, for the first Green Revolution. Today in the US, over 90% of the food and biofuel grain seeds are GE.

Transgenic seeds may be necessary for meeting the global food demand but they pose problems with equity, erosion, energy and ecological degradation. Transgenic seeds are no longer affordable to farmers in many countries, including small farmers in America. GE seeds consume more water and fertilizers while accelerating erosion because the seeds need extensive cultivation that compacts soil, decimates beneficial microorganisms, and degrades soil organics and structure.

Transgenic crops produce higher yields but cannot compete with natural grasses – weeds. Proof of their inability to compete with natural plants can be seen in the second year when a field swarms with weeds rather than volunteers from the transgenic crop. Consequently, farmers must till the soil before planting to remove weeds that would compete for soil moisture and nutrients. GE crops consume more energy because they receive additional cultivation and herbicides to control competing weeds and pests.

Highly productive seeds are designed for planting two or three times closer than traditional crops. Dense plantings consume substantially more water which means many dryland farmers have to put in irrigation systems. Dense plantings also consume significantly more fertilizer, which also creates more soil, air, and water pollution. Growing crops in crowded rows diminishes root development, degrades soil structure and yields produce without valuable micronutrients – vitamins, minerals, and antioxidants.

Transgenic seeds may offer modest relief in the sense that plants may be able to germinate in drier and saltier soils. However, little is known about the GE drought or salt tolerant seeds because they are still several years away and have not been tested in various growing situations. Transgenic seeds are typically more vulnerable to pests, weeds, and disease, which may offset other benefits.

Genetic engineering studies have shown that changing a target gene, to improve N absorption efficiency for example, also changes several unrelated genes that help protect the plant against disease. A gene change designed to improve tobacco's N efficiency changed the plants toxicological properties. Imagine what properties may change with genes that alter drought, heat, pH, and salt tolerance.

National Research Council reported that weeds are developing resistance to the herbicide that genetically engineered crops are designed to tolerate. Since genetically engineered crops were introduced in 1996, at least nine species of weeds have evolved resistance to glyphosate, a main component in Roundup™ and other commercial weed killers, largely due to repeated exposure.

Monsanto and other GM seed providers claim herbicide-resistant crops create less water pollution and minimize erosion. However, no infrastructure exists to track and analyze the effects that GE crops may have on erosion or water quality. Several independent studies have found just the opposite.

Many farmers, including those in the European Union find their communities and social belief systems forbid the use of "unnatural" seeds but GE material is almost ubiquitous in the global supply chain. About 30 GE varieties are grown now but by 2015 there are likely to be 120 GE varieties. Monsanto will soon market corn seed combining eight separate transgenic traits. Natural processes such as heritage seeds grown with biofertilizers may be necessary to provide sufficient affordable food in a manner that aligns with the prevailing social belief system in some crop-growing regions.

The rising cost of transgenic seeds magnifies social injustice; the difference between rich and poor farmers, communities and countries. The cost escalation for transgenic seeds occurs because a few companies practice oligopoly pricing. Three companies – BASF of Germany, Syngenta of Switzerland and Monsanto of St. Louis – have filed applications to control over two-thirds of the climate-related gene families submitted to patent offices worldwide. These climate ready genes may help crops survive drought, flooding, saltwater incursions, high temperatures, and increased ultraviolet radiation – which are predicted to undermine food security in coming decades.

Company officials deny the climate-ready seed applications amount to an intellectual property grab. They say GE seeds will be crucial to solving world hunger but would not be developed without patent protections. While that may be true, Monsanto has pushed hundreds of seed companies out of business and allows other seed companies to develop and patent seeds only if they use Monsanto's genes in their seed. Monsanto currently makes 60% of its revenue from GE seeds and does everything within its power to enforce, expand, and extend its GE franchise.

Some Monsanto seed prices increased 50% in 2008, upsetting farmers who must buy new seeds from the company every year. Monsanto has sued numerous farmers for saving seeds from their fields or using seeds the company said contained their patented genetic material. Critics such as competitor Pioneer Seeds calls Monsanto's behavior a "platform monopoly" that crushes competitors much like Microsoft's Window's platform.

Three varieties of Monsanto's genetically modified maize failed to produce crops during the 2008/9 growing season, leaving up to 200,000 hectares of corn fields barren of cobs across several provinces in South Africa. According the GRAIN SA, the varieties are: MON 810, NK 603 and MON 810 x NK 603. These seeds were sold to commercial maize farmers and provided for poor farmers in South Africa.

The *New York Times* reported that the Justice Department's antitrust division is investigating Monsanto for anticompetitive practices in the seed industry. The Justice Department is investigating whether Monsanto unfairly used genetic licenses to dominate the engineered seed market because 93% of soybean and 80% of corn plantings in the US in 2008 contained Monsanto's Roundup Ready™ trait. The Roundup Ready™ 1 trait patent expires in 2014 and Monsanto is forcing other seed suppliers to use their new Roundup Ready™ 2 trait in seeds that will effectively extend Monsanto's patent protection.

Britain's Soil Association study concluded that US GE crops have been an economic disaster which has caused some farm groups to call for a moratorium on GE wheat, the next proposed crop to be altered. The study estimated that gene-altered corn, soy, and rapeseed cost the US economy $12 billion since 1999 in farm subsidies, lower crop prices, loss of major export orders, and product recalls.

The biotechnology industry promised to solve the challenges of climate change and feeding the burgeoning world population while reducing agriculture's chemical impact. However, a 2009 study sponsored by the Union of Concerned Scientists examined the impacts of GE crops on pesticide use in the US and found a dramatic rise in the use of herbicides on genetically engineered crops. Charles Benbrook determined that 383 million additional pounds of herbicides have been used on GE crops since 1996, over what likely would have been used if GE crops had been replaced by conventional, non-GE varieties. The report shows the overall chemical footprint for engineered crops is immense and expanding. The growth in herbicide use has important implications for public health, the environment, and farmers' bottom lines.

The most contentious issue with GE crops is their impact on human and animal health. Everyone associated with transgenic crops knows that moving or adding a gene in a complex DNA sequence can have multiple unanticipated impacts on the organism. Monsanto did the R&D on their GE seeds and reported no adverse impact on mice feeding trials. For years, Monsanto held the research data privately, claiming intellectual property requirements.

A consortium of scientists from Greenpeace, Sweden and France recently won a judicial judgment that forced Monsanto to release the data from rat studies for three varieties of maize for scientific, third party examination. Unsurprisingly, the independent examiner found that the GE seeds did indeed impact the rodents' liver and kidneys. How much impact and toxicity remains under debate.

Each of the three maize strains caused unusual concentration of hormones and other compounds in the blood and urine of the tested rats, suggesting each strain impaired kidney and liver function. The rats that were fed the GE corn (NK 603) had elevated blood sugar levels and raised concentration of fatty substances called triglycerides. Both factors are potential precursors of diabetes.

Independent of GE seeds, many people assume additional technology innovations will rescue the world from the Malthusian calamity where population growth exceeds the food supply.

Technology rescue?

Sustainable food production may not be possible with traditional land-based crops unless we can undo the damage done from years of brutal chemotherapy to modern fields. Major innovations need to occur in several areas, Table 8.1.

Table 8.1 Innovations Needed for Crop Production

Innovation	Crop inputs to produce food
Nutrient efficient	Minimizes fertilizer application, nutrient waste and pollution.
Water light	Requires minimal or no fresh water.
Energy light	Requires minimal or no fossil energy.
Fertilizer light	Requires or no added chemical fertilizers
Chemical light	Requires minimal or no agricultural chemicals.
	Crop outputs – productivity
High yields	Grows high yields of good food on small acreage that diminishes the demand for cropland.
High quality	Grows produce with great taste, texture, and color.
High nutrient density	Grows produce with high density of macro and micronutrients, as well as vitamins and minerals.

Harvest waste	Harvestable with minimal waste.
Climate hardy	
Robust	Tolerant to crop stressors such as heat, wind, salt, insects, worms, weevils, and weeds.
Erosion resistant	Minimizes soil loss or degradation due to water and wind.
Drought tolerant	Germinates and produces with minimal water.
Salt tolerant	Thrives and produces in brine water and in soil invaded by irrigation, fertilizer or sea salt.
Ecosystem health – regenerative	
Builds soils	Adds nutrients and organics to the soil rather than constantly extracting.
Low tillage	Requires minimal or no tilling.
Perennial crops	Grows crops that do not have to be replanted every year.

Meaningful breakthroughs in these areas will take decades and cost significantly more than countries are willing to spend. Most these needed innovations for field crops lack even a theoretical model.

Chapter 9. Our Abundant Path Forward

You must be the change you want to see in the world.

– Mahatma Gandhi

Our world needs to develop a new approach to food production that does not rely on fossil resources because time is not on our side. Abundance with SAFE production offers solutions to many of the difficult challenges facing industrial agriculture and modern human societies. The rate of fossil resource consumption combined with perilously low reserves, makes the case for urgent action to develop a food and energy supply built on a foundation of plentiful and cheap resources. In spite of the obvious logic, there are strong headwinds on the path to abundance. The obstacles can be resolved, but will require creative solutions.

Humans have been traveling the land-based agriculture road for more than 11,000 years and will continue growing food crops in soil. Traditional agriculture takes its name from the habitual process of land-based crop cultivation and ingrained habits built on thousands of years of experience are very hard to break. Industrial agriculture enjoys intensive support worldwide that is unlikely to change – until the natural resources on which it depends become unaffordable or unaffordable. Prior to natural resource extinction, modern agriculture is highly vulnerable to significant food supply interruptions, which are likely to happen over the next 30 years, but could happen overnight.

Food supply interruptions

Our global food supply is closely interconnected. Supply shortages for food or the inputs for growing food in one region are likely to set off a food cascade that operates like a bank run. The perception of food scarcity motivates people to speculate, hoard, and steal food. Food shortages cause people to take horrific actions they would never consider if they and their family were not hungry. The food riots in 40 countries in 2008 killed and maimed thousands of people. Food riots may be a precursor to a food cascade. A small supply disruption can ignite market forces to create a firestorm that will kindle catastrophic physical and social damage as well hunger, malnutrition, disease, and deaths. No countries can afford the tremendous costs required to quell food riots while protecting food sources and supply chains.

The probability of each source of food supply interruption may be modest. However, taken together, the probability is probably greater than 80% that several will occur during the next 30 years. Then the only question under debate will be: "How could the experts have failed to anticipate either the events or the consequences?"

Consider the events that may create a 30% food crop loss to one or more large food growing regions. Food supply is highly inelastic. Small surpluses cause prices to plunge, while minor shortages or the anticipation of shortages produce abrupt price spikes.

Environmental source

- **Thirsty cities** – cities demand water as drought intensifies, leaving farmers without water.
- **Fierce storms** – destroys crops or prevents planting, cultivation, fertilization or harvest.
- **Dry winds** – vaporizes soil moisture and stresses, then kills crops.
- **Rain** – soaks the ground so machinery cannot harvest.
- **Floods** – destroy not only the crop but precious croplands.
- **Pollinator collapse** – widespread death to vital pollinators such as bees, bats, or butterflies.

- **Blight** – mildew, blight, rust, nematodes, or other disease vector destroys entire crops that are genetic monocultures.
- **Worms or insects** –attacked and destroy crops.
- **Pest resistance** – pests develop insecticide resistance. Farmers lose about 14% of their crops to pests today compared with 7% in 1950.
- **High heat** – unleashes a spectrum of fungal attacks that kill crops.
- **High morning temperatures** – blocks the plant from setting fruit.
- **High day temperatures** – stresses and then wilts the plants.
- **Prolonged drought** – stresses plants, devastates yields and consumes more water than is available.
- **Volcano** – blocks sun and creates 100% crop failure. (The Mount Tambora eruption of 1815 caused the summer without sun, where almost no crops grew in the US or Europe. Famine led to severe food riots, arson, and looting; then to the French Revolution.)
- **Aquifer depletion** – a major aquifer goes dry from over-extraction.
- **Wildfires** – drought leads to range fires that destroy cropland.
- **Rivers run dry** – rivers on which farmers depend for irrigation water go dry.
- **Irrigation salt invasion** – irrigation salts overwhelm cropland causing seeds to fail to germinate.
- **Ocean salt invasion** – rising oceans, tidal surges and storm surges destroy millions of acres of fertile cropland.
- **Environmental crash** – agricultural pollution causes an ecosystem to crash such as the Mississippi lowlands.
- **Hurricane** – destroys oil rigs and infrastructure, causing 50% price spikes in fossil fuels overnight.
- **Fierce ocean storm** -- disrupts shipping it prevents fuels and fertilizers from delivery to multiple countries.
- **Cyclone** – destroys farms, farm animals and the infrastructure for farming, including the cropland.

- **Mass migration** – a storm surge floods a heavily populated river delta, destroying infrastructure, and dumping salt in the aquifer, which forces people to move where they occupy cropland.

Human source

- **Death zone** – medical research confirms agricultural fertilizers and poisons in groundwater create not only a dead zone for aquatic species but disease and premature death for humans and animals.

- **Economic crisis** – insufficient funds for crop subsidies disrupt food production.

- **Trade war** – countries levy tariffs limiting trade in food and the critical inputs to grow food.

- **Crop insurance fails** – government runs out of money to pay farmers for failed crops due to widespread crop loss.

- **EPA** – begins enforcing fertilizer and poison run-off pollution.

- **Cities sue** – force farmers to stop well-water pollution.

- **Oil embargo** – oil crisis makes fossil fuels unavailable.

- **Mines crash** – phosphate or potash mines run out of product.

- **DNA threat.** Genetically engineered seed recall takes 50% of food grains seeds off the market. (Over 90% of the food grain farmers in the US currently plant GE seeds.)

- **GE seeds cause disease** – research confirms the suspicion that GE seeds increase risk for a set of diseases.

- **Fishermen sue** – force the agriculture industry to pay for their fishery loss to dead zones.

- **Seed supply** – production mistake fails to produce enough GE seed.

- **China** – experiences crop failures and buys the entire US grain production for a small fraction of the outstanding US debt.

- **Pirates** – close a critical shipping lane.

- **Terrorist strike on fuel or fertilizer** – destroy a major fuel refining facility or port that disrupts fuel supplies.

These food supply interruption scenarios illustrate that food shortages are highly likely to occur within the next 30 years. Building the capacity for abundance may not obviate food shortages but are likely to mitigate the severity. Modern agriculture is highly vulnerable to environmental or human caused events that cause substantial crop loss. Abundant agriculture can produce food reliably, independent of most external factors.

The path to abundant agriculture needs to start with public demonstration microfarms followed by diffusion to growers. Modern farmers are ideal abundance adopters because they bring a lifetime of experience in the husbandry of plants. Farmers are famously inventive and each will tweak their CAPS to maximize production in their local setting. Growers will continually upgrade their 360 microfarms and find new coproducts that benefit their farm, family, and community.

Production demonstration

Public demonstration of abundance and SAFE production will provide tremendous value because people could then visualize the simple process used to tap nature's oldest energy production system. Several cities are planning demonstrations sites that will provide research and education on SAFE production. Every city with a waste treatment plant, zoo, botanical gardens or aquarium offers great potential for co-locating abundance production systems to recycle wastes.

My favorite demonstration project, ZooPoo, would recycle and reuse the zoo animal and botanical wastes that are currently hauled off and burned or buried. The key metrics are phenomenal. ZooPoo contains 60% of the energy originally in the plant and over 80% of the nutrients. We can recycle these precious resources.

If we build an abundance CAPS on a hog farm outside of Oshkosh, a few dozen people year will see the system. When we build ZooPoo, millions of people will observe, engage, and learn about recycling human and animal waste as well as green energy systems.

Very few people have seen an algae production system in operation for three reasons:

1. Few outdoor algae production systems exist in the US and those producers are typically distant from cities.

2. Algae producers carefully guard their intellectual property and trade secrets and protect from pictures, notes and tours.

3. The federal government made a political decision to support corn ethanol research rather than algae in the 1990s, so there was no funding available for algae labs, training professors, graduate students, post docs or algae R&D for over 15 years.

Two indoor and outdoor algae labs and production facilities at Arizona State University provide a typical example. Both labs were built with over ten million dollars in state monies and hosted tours that excited students, faculty, and the community. Commercial enterprises made relatively small investments in the research and closed the labs and production facilities to the public – claiming intellectual property protection. The public investment that was promised to all is currently fenced to protect secrets.

Public access to information

Sufficient algae production knowledge exists to support abundance now. Unfortunately, much of the best knowledge rests in the brains of a select few scientists who must sequester their research findings due to intellectual property protections. Publicly funded projects should include a sunshine law – requiring public access to research findings.

The algae industry today is fractured into a series of vertical markets with each firm and public research facility acting as if they are a fiefdom where they must protect their intellectual property behind a wall of secrecy. Scientists are forbidden by non-disclosure agreements to collaborate or even talk honestly to other scientists about their projects. Their presentations at conferences are dumbed-down to protect their firm's intellectual property.

Scientists are forbidden to share their production breakthroughs, costs or productivity metrics. Researchers are embarrassed about their limitations in social situations and even informal conversations. Information concentration leads to mistakes in algae production that are repeated over and over again, which has been fatal for many emerging firms. Firms cannot admit or share their mistakes for fear that their next round of funding will dissolve. Concentrated knowledge severely limits innovation because breakthroughs depend on only a few brains rather than many.

Secrecy creates blindness and students, researchers and the press. Secrecy forces food and energy policy makers operate blindly to advances in algae production. The valuable information about algae production that students and entrepreneurs seek stays locked behind steel file drawers. Secrecy relegates students and the press to writing papers and articles on data that is 15 years old.

To facilitate public access, www.algaecompetition.com supports a collaboratory with open source architecture available to everyone. The knowledge base is searchable and organized under the relevant areas for algae production, harvest, component extraction, and coproducts. Published research, papers, ideas and experience contributed to the wiki will address the critical limitations and obstacles in order to improve algae production. A sticky wiki will include IP harvested from dated patents related to algae production.

The Sandra Day O'Connor College of Law at Arizona State University is creating an intellectual property pool for the algae industry where companies may contribute IP to the pool. Users will pay a small fee for use of the IP and additional fees based on production volume. Intellectual property pools have served several industries including chemicals, telecommunications and pharmaceuticals.

New technologies, especially in the areas of food and energy, are certain to create unanticipated and unintended consequences. Active scientific, social, and political analysis of SAFE production will be important to minimize unintended consequences. Probably the strongest fear for algaculture is that it would host some form of disease such as salmonella, e-coli or other pathogen.

The key solution for assuring positive impacts will be creating metrics, monitoring production and reporting in a manner that assures the public, scientists and policymakers have access to the data. Great

Britain has developed a set of food safety metrics that may provide a model for monitoring and reporting the quality of abundance production.

Supply and Demand

Production advances in the food supply system, especially those that are regenerative versus extractive, will improve food security. However, meeting the nutritional needs of current and future generations will require far more than additional food production. Actions that need to be taken include the following.

1. **Population management.** Countries and communities must find ways to slow population expansion. Our planet simply lacks the natural resources for food, water, shelter and transportation for many more people. We are currently over-consuming the Earth's natural resources by over 30%.

2. **Waste management.** Roughly half of our food is wasted in the field and along the supply chain. In the US, many consumers waste half of the food they buy. We need to educate farmers, food marketers, and consumers to find ways to minimize food, water and energy waste.

3. **Consumer behavior.** We need to reverse the trend of higher consumption of resource-expensive meats and dairy products. Food prices should reflect their ecological and resource impact. Each pound of meat consumes 18 times more fresh water than food grains, yet neither farmers nor consumers pay for resource depletion. Affordable food will increase as consumers eat less meat and dairy and more grains, fruits, and vegetables.

4. **Recognize and reward food production innovation.** Every community, region and nation celebrates their best chefs. Why not recognize and reward for sustainable food producers for resource preservation and pollution avoidance?

5. **Produce locally.** Escalating energy costs will make long-distance food transport a distant memory soon. We need to develop local production models that grow food in and near large cities as well

as every community. Vertical farms and microfarms can provide strong food production platforms.

Abundance methods can play a critical role in assuring plentiful good food for our children and our global neighbors.

Food security

Food security should be near the top among national security priorities. While America has focused on terrorism and immigration, the silent and invisible hand of fossil resource extinction paired with global climate change threatens our society. Failing food security, all social systems degrade with the serious threat of social destruction. Food security requires some practical actions.

- **Appoint a Chief Sustainability Officer** in the Office of Homeland Security with the charge to monitor and apply best practices to manage and conserve the precious fossil resources on which our food supply depends. This office should also have responsibility to track and quantify the costs of fossil resource extraction and agricultural and industrial erosion and pollution.
- **End subsidies and protective tariffs** for unsustainable and ecological destructive production such as corn ethanol. Subsidies should be shifted to sustainable and ecologically positive production of food, fibers and possibly biofuels.
- Create a **Plant Microbiome Project** similar to the National Institute of Health's Human Microbiome Project to identify and genetically sequence the different microbial genera.
- Develop **statutory soil protection policies** analogous to those applied to air and water.
- Recognize and designate **algae production as farming**. Currently, the USDA does not recognize algae production as farming or aquaculture which means algae farmers cannot obtain USDA or Farm Credit Services loans.
- **Build ZooPoo**, a nutrient recovery exhibit at one zoo, botanical garden, or science center and then build others.
- **Create a nano aquarium** that enables guests and researchers to see the incredible actions of microbial biofactories as they manufacture proteins, vitamins, and antioxidants.

- Develop and give public access to an **abundance metrics monitoring system** for sustainability, avoidance of synthetic or mined chemicals, ecological indicators, yields, size, color, soil structure, root size and length. Monitor water and energy use, crop productivity improvement; produce quality enhancements, biological diversity, fertilizer use and type, avoidance of chemical poisons and pollution mitigation.
- **Create a sustainability label** for products that indicates ecological, economic and social impacts.
- Create a **national breeding program** for high yield, high nutrient produce that halts nutrient erosion in our foods.

Additional government commitment may:

- Invest in abundance **R3D** (R&D, demonstration, and diffusion).
- Build and operate at least one **360 microfarm** demonstration unit in every state, and then in every city.
- Create **standards** for algae seed culture quality, which may be an extension of existing organic standards.

Farmer education in abundance practices may begin with the knowledge transfer from agricultural research universities.

- **Train the trainers** – retired farmers, master gardeners and college students who will train microfarmers.
- **Build a website** with training and support materials.
- **Create social networks** to support knowledge transfer, best practice, demonstration and diffusion.
- **Teach sustainable agriculture** in high schools, institutes, universities, and colleges.

Engage organic gardening, permaculture, botanic gardens, zoos, aquariums, science centers, high schools colleges, and universities in abundance R3D.

Create competitions and rewards for abundance innovations including the most microbially manufactured N, most efficient nutrient recovery system, best nutrient-delivery model, most efficient irrigation system, and other sustainable farming outcomes.

Build collaborative R3D with green organizations such as the Sierra Club, the Rodale Institute, World Wildlife Fund, Greenpeace, The Heifer Project, Rotary International and others.

Your ideas will move abundance and sustainable food production forward. Please share your insights, ideas, inventions, musings, aha's, collaborations, and actions at www.AlgaeCompetition.com.

Our Shared Path Forward

Our children's future depends on food security. Two critical threats to supplying sufficient food are the mass extinction of fossil resources and global climate change. As the price of fossil fuel rises and the availability of agricultural chemicals diminishes, farmers will need to find sustainable inputs that rely on neither fossil energy nor extracted minerals. Our current food supply will be decimated by the onset of fierce storms, hotter and more acidic oceans, rising sea levels that devastate cropland and cause mass migrations, higher temperatures, hot dry winds, prolonged droughts, and massive wildfires.

Human societies currently exploit roughly half of the total primary biological production of the planet – through production of food, fisheries, forests, fabrics, medicines, and other activities. Producers have modified, manipulated, or mined natural ecosystems on all continents and oceans have been to support human consumption.

Industrial agriculture has increased food production and enabled significant population expansion that now depends on fossil resources which are rapidly diminishing in supply and increasing in cost. Modern agriculture has accelerated the extraction of non-renewable fossil resources in a manner that is unsustainable. Global ecosystem exploitation will collapse when the first of 21 critical fossil resources becomes unaffordable or unavailable. Human societies will survive or perish in the near future based on our ability to manage the Earth's remaining biological and physical assets sustainably.

Nature created abundance, a SAFE production system that produces without reliance on fossil resources. Abundance is largely immune to the complex impacts on food production imposed by climate change.

Abundant Agriculture

Abundance farmers using 360 microfarms can produce food and coproducts for their family and community anywhere on Earth.

Farmers will embrace the opportunity to transform the costs associated with their waste stream to a profit center with the 360 microfarms. Society will celebrate the process of recovering energy and nutrients from waste streams, while reversing air, soil, and water pollution.

Additional research needs to focus on every abundance element. Growers will need training in the design, development and operations involved in using algae to remediate their waste streams and deliver nutrients to their crops. New industries will emerge to facilitate the use of algae to clean and recycle nutrients from waste streams and to produce food and other less critical forms of energy, such as liquid transportation fuels.

Algae are available all over the Earth and are prepared to do their ingenious work to support microfarmers, hungry consumers, and society. We must act quickly to take advantage of algae's green promise – to create food security for everyone on our planet. Together, we can endow our children with a superb legacy never before achieved – sustained abundance.

Appendix I. Soil Macro and Microorganisms

Element	Function in plants
Actino-mycetes	Actinomycetes (ac"-ti-no-my'-cetes) are thread-like bacteria that look like fungi. While not as numerous as bacteria, they perform vital roles in the soil. They help decompose organic matter into humus which slowly releases nutrients. They also produce antibiotics to fight roots diseases. The same antibiotics are used to treat human diseases. Actinomycetes create the sweet, earthy smell of biologically active soil when a field is tilled.
Algae	Thousands of algal species live in the top six inches of soil. Unlike most other soil organisms, algae produce their own food through photosynthesis. However, if there is not enough light, they can use organic nutrients available in the soil. They appear as a green or multicolored film on the soil surface following rain or irrigation. Algae improve soil structure by producing substances that glue soil together into water-stable aggregates. Some species of algae (the blue-greens) can fix their own N, some of which is later released to plant roots.
Arthropods	Arthropods are invertebrate organisms having an external skeleton, a segmented body and jointed appendages. Over a million species have been described in the science literature and they make up more than 80% of all known living species. They range in size from microscopic plankton up to forms a few meters long. Arthropods include the insects, arachnids and crustaceans. Common soil arthropods include sowbugs, millipedes, centipedes, slugs, snails and springtails.

	Arthropods, along with worms, are the primary decomposers. They eat, shred plant and poop animal residues. Some bury residue, allowing other soil organisms to further decompose it. Springtails are a small insect that eat mostly fungi. Their waste is rich in plant nutrients that are released after other fungi and bacteria decompose it. Dung beetles play a valuable role in recycling manure and reducing livestock intestinal parasites and flies.
Bacteria	Most numerous among soil organisms are the bacteria; every gram of soil contains at least a million of these tiny one-celled organisms. There are many different species of bacteria, each with its own role in the soil ecosystem. Bacteria break down complex molecules and enable plants to take up nutrients. Some species release N, S, P and trace elements from organic matter. Others break down soil minerals and release K, P, Mg, Ca and Fe. Other species make and release natural plant growth hormones which stimulate root growth.\n\nA few bacteria fix N in the roots of legumes while others fix N independently of plant association. Bacteria are responsible for converting N from ammonium to nitrate and back again depending on soil conditions. Various bacteria species increase the solubility of nutrients, improve soil structure, fight root diseases, and detoxify soil.
Earthworms	Earthworm burrows enhance water infiltration and soil aeration. Earthworm tunneling can increase the rate of water entry into the ground 4 to 10 times higher than fields that lack worm tunnels.[1] This reduces water runoff, recharges groundwater, and helps store more soil water for dry spells. Vertical earthworm burrows pipe air deeper into the soil, stimulating microbial nutrient cycling at those

deeper levels. Tillage done by earthworms can replace some expensive tillage work done by machinery.

Worms eat dead plant material left on top of the soil and redistribute the organic matter and nutrients throughout the topsoil layer. Nutrient-rich organic compounds line the tunnels that may remain in place for years if not disturbed. During droughts these tunnels allow for deep plant root penetration into subsoil regions of higher moisture content. In addition to organic matter, worms also consume soil and soil microbes as they move through the soil. The soil clusters they expel from their digestive tracts are known as a worm castings. Each worm cast is separate from other casts and ranges in size from that of a mustard seed to a sorghum seed depending on the size of the worm. The soluble nutrient content of worm casts is considerably higher than those of the original soil (see Table 2). A good population of earthworms can process 20,000 pounds of topsoil per year, with turnover rates as high as 200 tons per acre having been reported in some exceptional cases.[2]

Fungi

Fungi come in many different species, sizes and shapes in soil. Some species appear as thread-like colonies, while others are one-celled yeasts. Slime molds and mushrooms are also fungi. Many fungi aid plants by breaking down organic matter or by releasing nutrients from soil minerals. Fungi are generally early to colonize larger pieces of organic matter and begin the decomposition process. Some fungi produce plant hormones, while others produce antibiotics including penicillin. Several fungi species trap harmful plant-parasitic nematodes.

Mycorrhizae	The mycorrhizae (my-cor-ry-'zee) group of fungi lives either on or in plant roots and act to extend the reach of root hairs into the soil. Mycorrhizae increase the uptake of water and nutrients especially in less fertile soils. Roots colonized by mycorrihizae are less likely to be penetrated by root-feeding nematodes since the pest cannot pierce the thick fungal network. Mycorrhizae also produce hormones and antibiotics, which enhance root growth and provide disease suppression. The fungi benefit from plant association by taking nutrients and carbohydrates from the plant roots where they live.
Nematodes	Nematodes are abundant in most soils and eat decaying plant litter, bacteria, fungi, algae, protozoa and other nematodes and speed the rate of nutrient cycling. A few species are harmful to plants.
Protozoa	Protozoa are free-living microorganisms such as amoeba that crawl or swim in the water between soil particles. Soil protozoa are predatory and feast on other microbes, including bacteria. Protozoa accelerate the cycling of N from the bacteria, making it more available to plants.

Adapted from: Sullivan, Preston. Appropriate Technology Transfer for Rural Areas, (ATTRA).

http://www.soilandhealth.org/01aglibrary/010117attrasoilmanual/0101 17attra.html#2

Acknowledgements

Thanks to Ann who brings abundance to my life every day. Ann created fabulous gourmet algae dinners and wonderful support, even when friends considered my algae R&D odd and geeky.

Abundance would not have been possible without the extraordinary research of David and Marcia Pimentel, Lester Brown, President of the Earth Policy Institute, Jeffery Sachs of the Earth Institute. Professors Qiang Hu, Milton Sommerfeld and Bruce Rittman supported endless questions on molecular biology and algae production. Environmental scientists Al Darzins, Eric Jarvis and Mike Siebert at NREL were very helpful with renewable energy sources. Thanks also to great advisors who elevated *Abundant Agriculture* from a solo to an orchestra.

Science	Business – Econ.	Agribusiness
Robert Henrikson	Lucian Spataro	Jon Ewen
Dan Childers	Mark Allen	Gary Wood
James Elser	Alan Resnik	Doug Young
Jessica Corman	Gary Dyer	Jim Robertson
Ben Cloud	Herb Roskind	Tracy Penwell
Andy Ayers	William Cockayne	Barry Spiker

Also helpful were the published works of Paul Ehrlich, Sandra Postel, Nobel Laureate Al Gore, Harvey Blatt, Fred Pearce, Michael Pollen, Brian Halweil, Clay Jason and Linda Graham. High-content websites were a great support such as Algaebase, U.N., W.H.O., the National Resources Defense Council, Sierra Club, Green Peace, Audubon Society, Union of Concerned Scientists, Center for Energy and Climate Solutions, Clean Water Network and Public Citizen. Also useful were US government sources including: DOE, EPA, USDA, NOAA and NREL.

Mark Edwards

Mark engineers nutritious, sustainable and affordable food and energy (SAFE) production that is available to everyone on Earth. Mark pursues abundance; to create food security for all and help all growers leave every field better than they found it.

Hunger, nutrition, pollution and soil regeneration are urgent global challenges. Abundant agriculture and the use of smartcultures provide a practical path for improving our nutrition, lowering food costs and ending ecological pollution. We need to engage millions of Green Masterminds globally who have the capacity to grow algae as food, feed and fertilizer for the needs of their family and community.

Mark graduated from the U.S. Naval Academy in mechanical engineering, oceanography and meteorology where Jacques Cousteau motivated and mentored his interest in the oceans and global stewardship. He holds an MBA and PhD in marketing and consumer behavior and has taught agribusiness food marketing, sustainability and entrepreneurship at Arizona State University for over 30 years.

Mark served as CEO of TEAMS Intl. for 24 years, the software and assessment firm he founded based on his research on advanced assessment technologies, talent and leadership assessment. He served as lead consultant for more than 400 firms globally. He was retained by many U.S. departments and the military, including DOE, DOD, Special Forces and the National Labs.

Mark served as a Director for a Fortune 50 foods company and has done extensive R&D on new foods, sources and consumer behavior. He has consulted for Monsanto, Pioneer Seeds, DuPont, Nabisco, Quaker Oats, General Mills, Borden and many other agribusiness companies. He has worked with senior executives at 15 large U.S. oil and gas firms as well as British Petroleum and Saudi Aramco.

Great Green Reading

Feed our World

Smil, Vaclav. *Feeding the World: A Challenge for the Twenty-First Century*, The MIT Press, 2001.

Winne, Mark. *Closing the Food Gap: Resetting the Table in the Land of Plenty*, Beacon Press, 2009.

Patel, Raj. Stuffed and Starved: The Hidden Battle for the World Food System, Melville House, 2008.

Lyson, Thomas A. *Civic Agriculture: Reconnecting Farm, Food, and Community*, Tufts, 2004.

Hinrichs, C. Clare and Thomas A. Lyson, *Remaking the North American Food System: Strategies for Sustainability*, University of Nebraska Press, 2009.

Federico, Giovanni. *Feeding the World: An Economic History of Agriculture, 1800-2000*, Princeton University Press, 2008.

Edwards, Mark R. *Crash! The Demise of fossil Foods and the Rise of Abundance*, Tempe: CreateSpace, 2009.

Conway, Gordon. The doubly green revolution: food for all of the 21st century, Ithaca: Cornell University Press, 1997.

Sustainable Agriculture and Permaculture

Andersen, Arden B. Science in Agriculture: Advanced Methods for Sustainable Farming, Acres USA., 2000.

Gliessman, Stephen R. Agroecology: The Ecology of Sustainable Food Systems, Second Edition, CRC Press; 2 ed., 2006.

John Ikerd. Crisis and Opportunity: Sustainability in American Agriculture, Bison Books. 2008.

Mason, John. *Sustainable Agriculture*, 2nd Ed, Landlinks Press, 2003.

Clay, Jason. World agriculture and the environment, Washington: Island Press, 2004.

Soils

Sandra Postel, *Pillar of Sand: Can the Irrigation Miracle Last?* W. W. Norton & Company, 1999.

Magdoff, Fred and Ray R. Weil. *Soil Organic Matter in Sustainable Agriculture* (Advances in Agroecology), New York: CRC Press, 2004.

Daudu, Christogonu.s Organic Matter Sources, Soil Fertility and Productivity, VDM Verlag, 2008.

Plaster, Edward J. Soil Science and Management, 5th. Ed., Delmar Cengage Learning; 2008.

Stengel, P and S. Gelin, Eds. *Soil: Fragile Interface*. Science Publishers, Plymouth, UK, 2003.

Food, energy and economics

Thomas L. Friedman, *Hot, Flat, and Crowded: Why We Need a Green Revolution – and How It Can Renew America*, Farrar and Giroux, 2008.

Lester R. Brown, *Plan B 4.0: Mobilizing to Save Civilization*, 4th Ed., W. W. Norton; 2009.

Jeffrey D. Sachs, *Common Wealth: Economics for a Crowded Planet,* Penguin Press HC, 2008.

Smil, Vaclav. *Global Catastrophes and Trends: The Next Fifty Years*, The MIT Press, 2008.

Water

Elizabeth Kolbert, *Field Notes from a Catastrophe: Man, Nature, and Climate Change*, Bloomsbury, 2006.

Peter H. Gleick, *The World's Water 2006-2007*: *The Biennial Report on Freshwater Resources*, Island Press, 2006.

Fred Pearce, *When the Rivers Run Dry: Water – The Defining Crisis of the Twenty-first Century*, Beacon Press, 2007.

Glennon, Robert Jerome. *Unquenchable: America's Water Crisis and What to Do About It*, Island Press, 2009.

Our Abundant Path Forward

_____

www.ingramcontent.com/pod-product-compliance
Lightning Source LLC
Chambersburg PA
CBHW051516170526
45165CB00002B/496